THE TROPICAL AGRICULTURIST

*Series Editor*
René Coste
Formerly President of the IRCC



Pi

Tr

CTA
MACMILLAN

Original French edition published in 1987 under the title *Le Manioc* in the series *Le Technicien d'Agriculture Tropicale,* by the ACCT and Maisonneuve et Larose, 15, rue Victor-Cousin, F75005 Paris, France.
No responsibility is taken by the holders of the copyright for any changes to the original French text.

This edition published 1989 by
MACMILLAN EDUCATION LTD
London and Basingstoke
*Associated companies and representatives in Accra, Banjul, Cairo, Dar es Salaam, Delhi, Freetown, Gaborone, Harare, Hong Kong, Johannesburg, Kampala, Lagos, Lahore, Lusaka, Mexico City, Nairobi, São Paulo, Tokyo*
Published in cooperation with the Technical Centre for Agriculture and Rural Cooperation, P.O.B. 380, 6700 AJ Wageningen, The Netherlands.

ISBN 0-333-47395-7

11 10 9 8 7 6 5 4 3 2
06 05 04 03 02 01 00 99 98 97

Printed in Hong Kong

A catalogue record for this book is available from the British Library.

The opinions expressed in this document and the spellings of proper names and territorial boundaries contained therein are solely the responsibility of the authors and in no way involve the official position or the liability of the Agency for Cultural and Technical Cooperation or the Technical Centre for Agricultural and Rural Cooperation.

The ACP-EU Technical Centre for Agricultural and Rural Cooperation (CTA) operates under the Lomé Convention between Member States of the European Union and the African, Caribbean and Pacific States.

CTA collects, disseminates and facilitates the exchange of information on research, training and innovations in the spheres of agriculture and rural development and extension for the benefit of the ACP states.

To achieve this, CTA commissions and publishes studies, organises and supports conferences, workshops and seminars, publishes and co-publishes a wide range of books, proceedings, bibliographies and directories, strengthens documentation services in ACP countries and offers an extensive information service.

Postal address: Postbus 380, 6700 AJ Wageningen, The Netherlands.

*Agency for Cultural and Technical Cooperation (ACCT)*

The Agency for Cultural and Technical Cooperation, an intergovernmental organisation set up by the Treaty of Niamey in March 1970, is an association of countries linked by their common usage of the French language, for the purposes of cooperation in the fields of education, culture, science and technology and, more generally, in all matters which contribute to the development of its Member States and to bringing peoples closer together.

The Agency's activities in the fields of scientific and technical cooperation for development are directed primarily towards the preparation, dissemination and exchange of scientific and technical information, drawing up an inventory of and exploiting natural resources, and the socio-economic advancement of young people and rural communities.

*Member countries:* Belgium, Benin, Burkina Faso, Burundi, Canada, Central African Republic, Chad, Comoros, Congo, Côte d'Ivoire, Djibouti, Dominica, France, Gabon, Guinea, Haiti, Lebanon, Luxembourg, Mali, Mauritius, Monaco, Niger, Rwanda, Senegal, Seychelles, Togo, Tunisia, Vanuatu, Vietnam, Zaire.

*Associated states:* Cameroon, Egypt, Guinea-Bissau, Laos, Mauritania, Morocco, St Lucia.

*Participating governments:* New Brunswick, Quebec.

**Titles in *The Tropical Agriculturalist* series**

| | | | |
|---|---|---|---|
| Sheep | ISBN 0-333-52310-5 | Ruminant Nutrition | 0-333-57073-1 |
| Pigs | 0-333-52308-3 | Warm-water Crustaceans | 0-333-57462-1 |
| Goats | 0-333-52309-1 | Animal Breeding | 0-333-57907-0 |
| Dairying | 0-333-52313-X | Animal HealthVol.1 | 0-333-61202-7 |
| Poultry | 0-333-52306-7 | Animal Health Vol.2 | 0-333-57360-9 |
| Rabbits | 0-333-52311-3 | Livestock Production Systems | 0-333-60012-6 |
| Draught Animals | 0-333-52307-5 | | |

| | | | |
|---|---|---|---|
| Upland Rice | 0-333-44889-8 | Sugar Cane | 0-333-57075-8 |
| Tea | 0-333-54450-1 | Maize | 0-333-44404-3 |
| Cotton | 0-333-47280-2 | Plantain Bananas | 0-333-44813-8 |
| Weed Control | 0-333-54449-8 | Coffee Growing | 0-333-54451-X |
| Spice Plants | 0-333-57460-5 | Food Legumes | 0-333-53850-1 |
| Cocoa | 0-333-57076-6 | Cassava | 0-333-47395-7 |
| Storage of Food Grains and Seeds | 0-333-44827-8 | Sorghum | 0-333-54452-8 |
| Avocado | 0-333-57468-0 | Cut Flowers | 0-333-62528-5 |

**Other titles published by Macmillan with CTA** ***(co-published in French by Maisonneuve et Larose)***

| | |
|---|---|
| Animal Production in the Tropics and Subtropics | 0-333-53818-8 |
| Coffee: The Plant and the Product | 0-333-57296-3 |
| The Tropical Vegetable Garden | 0-333-57077-4 |
| Controlling Crop Pests and Diseases | 0-333-57216-5 |
| Dryland Farming in Africa | 0-333-47654-9 |
| The Yam | 0-333-57456-7 |

**The Land and Life series** ***(co-published with Terres et Vie)***

| | |
|---|---|
| African Gardens and Orchards | 0-333-49076-2 |
| Vanishing Land and Water | 0-333-44597-X |
| Ways of Water | 0-333-57078-2 |
| Agriculture in African Rural Communities | 0-333-44595-3 |

# Contents

# 1 Basic data

## 1.1. *Cassava worldwide*

### Production

Some 14 million hectares of cassava (or 'manioc', as it is also known) are grown worldwide, producing 120 to 130 million tonnes of tubers. The area of land planted with cassava is greatest in Africa, but yields there are also lower than anywhere else.

**Table 1**

| | **Area** ('000 ha) | **Production** ('000 t) | **Yield** (t/ha) |
|---|---|---|---|
| **South America** | 2 800 | 30 000 | 11 |
| **Asia** | 4 000 | 45 000 | 11 |
| **Africa** | 8 000 | 48 000 | 6 |

Fifteen countries produce in excess of a million tonnes. Of these, the five biggest producers — Brazil, Indonesia, Nigeria, Thailand and Zaire — each produce more than 10 million tonnes, and together account for 63% of world output.

Average yields in the producer countries are not very high. The following examples may be given (t/ha):

**Thailand:** 13 to 17 **India:** 15 to 18 **Angola:** 14 to 15
**Brazil:** 10 to 11 **Ghana:** 7 to 8 **Zaire:** 7

### Uses

Cassava is generally used directly for human consumption in the producer countries where it is processed on a small scale in a rather unsophisticated manner. The sweet varieties are simply boiled, but in the case of the bitter varieties, the cassava is prepared in various ways, to make *gari*, *atieké*, *chikwangue* or *fufu*, as the dishes are known in Africa, and *farinha de raspa or kwako*, *farinha de agua*, etc., in South America where cassava originated.

Cassava is also used in livestock feed. In the producer countries, the peelings or various forms of dried cassava (chips or meal) are fed to animals. However, it is in the non-producer countries, particularly in Europe, where there has been a considerable increase in the use of cassava as an animal feedstuff. Virtually all the cassava traded internationally is used for this purpose.

International trade in dried cassava, in the form of meal or more generally pellets, accounts for 10 to 15% of world production. This trade may be summarized as follows (1984 figures):

| | | |
|---|---|---|
| *Imports* | in '000 tonnes of total fresh root equivalent: | 14 500 |
| | of which Western Europe (EEC): | 12 200 |
| *Exports* | in '000 tonnes of total fresh root equivalent: | 14 500 |
| | of which Thailand: | 13 600 |

As can be seen, the bulk of world trade in dried cassava is conducted between Thailand and Europe. The other exporting countries are Indonesia, whose share of world trade is tending to decline, and certain Latin American countries, which account for only small amounts. The other importers are increasingly Asian, particularly China.

The other main cassava products which are traded internationally are starch and processed starches such as tapioca, but these account for only 1 to 2% of world production.

As the producer countries develop economically, they are increasingly likely to industrialize their small-scale processing operations in order to supply their own domestic markets with high-quality foodstuffs, and also to meet export demands.

Economic development gives rise to new consumers and a demand for new products; it also tends to lead to a reduction in the number of producers and obliges them to increase productivity. Farmers must produce more, without increasing their labour input. To achieve that, they must farm more efficiently, use more effective techniques and be well informed regarding those techniques. This book hopes to contribute towards providing the necessary information.

## 1.2. Cassava-based farming systems

Cassava is grown in farming systems which may fall into three main categories:

(a) shifting cultivation systems
(b) 'pioneer' systems
(c) permanent systems

## Shifting cultivation

The term 'shifting cultivation' applies to a system in which the fields are cultivated for just a few years and then abandoned, allowing the natural vegetation to take over for a period well in excess of that for which the land has been cropped. Normally, a new plot is opened up for cultivation every year by clearing an area which has never been farmed before, or which has been left fallow for a long while.

Shifting cultivation is common where there is plenty of land available in relation to the human population, and where modern inputs such as machinery, fertilizer or pesticides cannot be deployed. In the absence of

Fig 1 *Growing cassava the traditional way, Indonesia*

these modern techniques, it is possible with shifting cultivation to manage the soil's natural fertility fairly well, control weeds easily and, in some cases, curtail the effects of parasites.

This type of system is always extensive (production per unit of land used is low) and the plots are worked by hand. The bulk of world cassava production is still grown under shifting cultivation systems. The practices still followed by certain Indians in the Amazon basin, where production methods appear to have remained unchanged since time immemorial, provide one of the most typical examples.

In the Amazon basin, which is probably one of the centres where cassava originated, each Indian family clears about one hectare of forest every year, of which approximately half is cultivated, given that the edges and any unsuitable areas are left untouched. Felling and burning during and at the end of the dry season are the most important operations. The soil is loosened with a stick in those places where the cuttings are to be planted, at a spacing of about 1.5 m × 1.5 m Two or three cuttings 20 to 50 cm long are planted together at a time. Harvesting is spread between the 10th and 22nd months after planting, and is done every fortnight. Under this system:

- a very large number of cassava varieties are grown together;
- cassava is interplanted with other crops, such as maize, sugar cane, bananas, etc., but these account for less than 1% of the total area under cultivation;
- fields are generally cropped once only.

If the same crop is grown twice in succession on the same field, this will increase the amount of weed control required (which was previously virtually non-existent), and exacerbate the damage done by cassava ants. Yields may become very low.

In Africa too, a large proportion of the cassava is produced under shifting cultivation systems. Usually, after the shrubby bush or savanna has been cleared, the fields are cropped for one to three years, a combination of cereals, vegetables and other species being grown, and cassava cuttings being interplanted with the last crop of the cycle. During the first few months the cassava benefits from the attention given to that crop, after which it is left alone. Harvesting is phased according to requirements, and generally takes place between ten months and two years after planting.

Two examples are given below of cassava being grown in Africa under a traditional shifting cultivation system.

In Zaire, in the savanna areas, farmers often operate two different rotation systems, each ending with cassava, prior to the land being laid fallow again. Preparations for one of the rotations involve clearing the land and slowly burning the natural vegetation, after burying it under wide ridges, the first crops being demanding ones. The other rotation

Fig 2 *Field surrounded by thorn bushes in Burkina Faso*

involves preparing the land by clearing and simply burning off the vegetation, and starts with hardier crops. Generally speaking, the second rotation is shorter than the first.

The crop sequences are as follows:

| FIRST YEAR | 1st rotation | 2nd rotation |
|---|---|---|
| **dry season** | clearing the land and slowly burning off vegetation under ridges | clearing the land and simply burning off vegetation |
| **1st wet season** | maize, beans<br>yams, *Cucurbitaceae* | groundnuts or<br>groundnuts/cassava |
| **2nd wet season** | yams (following on from 1st season) | cassava (following on from 1st season) or groundnuts, cassava (if no cassava planted during 1st season) |

| SECOND YEAR | 1st rotation | 2nd rotation |
|---|---|---|
| **dry season** | weeding and simply burning vegetation | cassava (continued) |

| | | |
|---|---|---|
| **1st wet season** | groundnuts or groundnuts/cassava This rotation then continues as for the first year (2nd rotation) | cassava (continued) harvesting as and when required; land reverts to fallow |

On average, there are about six people on each smallholding, 2.5 of whom are active, and 1.5 ha of land is cropped.

In the Central African Republic (CAR) in the cotton-growing regions, the crop rotation is usually as follows:

| | |
|---|---|
| **FIRST YEAR** | |
| **dry season** | clearing the savanna by simply burning off the vegetation |
| **rainy season** | cotton |

| | |
|---|---|
| **SECOND YEAR** | |
| **dry season** | clearing the ground by burning |
| **rainy season** | sorghum, maize, groundnuts grown over one or two crop cycles, certain species being intercropped — cassava cuttings planted in first or second cycle |

| | |
|---|---|
| **THIRD YEAR** | cassava, harvesting and return to fallow |

Under a shifting cultivation system, cassava yields are generally low (of the order of 10 t/ha), but the amount of work required is also low — about ten days per tonne of roots harvested and transported. Shifting cultivation can continue to be practised only in those areas where land is still plentiful. As land becomes scarce, the length of the fallow period is shortened, the soil is impoverished, annual weeds take over and yields decline. This means more work for a smaller crop. In such situations, new techniques must be used and production intensified.

## Pioneer system

This is a system for agricultural development rather than a system of production.

Whenever virgin land is being developed by the planting of perennial species, food or cash requirements prior to the plantation's coming on stream often lead farmers to intercrop annual or quick-growing species during the early years.

This is how cassava-growing first developed in Thailand. There, it was first grown as an intercrop in young rubber-tree plantations. This method of production is found in a number of other countries (such as Brazil, Indonesia, Ivory Coast), where cassava is often planted for industrial purposes or as a food crop in combination with young rubber, coffee or cocoa trees.

Such a system is practised by both smallholders — which is generally the case in Africa in the coffee and cocoa plantations — and large scale farms. In the former case, the cassava is grown manually; in the latter, operations may be partially mechanized, particularly the preparatory and, where appropriate, maintenance work.

## Permanent systems

In order to grow cassava under permanent systems, a number of conditions must ideally be met:

- techniques must be applied to conserve the soil and maintain soil fertility, such as measures to combat erosion, the use of organic and/or artificial fertilizers, rotation with cover plants (green manuring or grazed forage);
- weeds must be controlled;
- operations must to some extent be mechanized, this often being made necessary by the techniques used (ploughing up fallow land, tilling the soil) and for economic reasons (reduced labour costs in order to offset the cost of inputs).

These requirements are not always met. Permanent systems under which cassava is grown may be classified according to the degree of mechanization and modernization.

### Manual cultivation

There are two types of permanent system under which cassava is grown manually. One results from a debased form of shifting cultivation, the other involves growing cassava alongside vegetables in intensive market gardens close to urban centres.

An example of shifting cultivation that has gradually lapsed into a permanent system is to be found in certain overpopulated regions of Togo, where plots which used to be left fallow for long periods are now cropped continuously with maize. This crop which is very demanding has gradually given way to cassava, which is hardier and is now often grown continuously alone. The techniques used to preserve soil fertility (mainly low applications of organic domestic waste and interplanting or alternating with a legume such as cowpea — *Vigna*) are inadequate. Yields are very low. The fact that virtually the entire crop is consumed by the farmer and his family makes it difficult to purchase modern inputs.

In contrast, the market gardens operate a highly intensive permanent farming system, offering the benefits of considerable amounts of fertilizer and of irrigation. While it is still only a modest sideline, cassava is increasing in importance in this type of system, particularly in the dry regions of Africa where it is valuable as a staple 'energy' food.

### Cultivation using draught animals

In Africa, draught animals are hardly used at all in growing cassava, for in the regions that are most suitable for growing this crop there are no such animals available because of trypanosomiasis. However cassava is also important in seasonally dry areas where cattle can be kept.

Draught animals are used in growing cassava in Thailand, but there they are tending to be replaced by tractors. In Thailand, the average size of the farms where draught animals are used is 7 ha, of which 4 ha are planted with cassava. Oxen are used for preparing the land — which is done very carefully, the land often being ploughed twice or even three times, followed by harrowing — and for ridging, which is done after the cuttings have been planted by hand. Under this farming system, artificial fertilizer and herbicides are frequently used.

### Partially mechanized cultivation

Operations were partially mechanized in Madagascar on farms which used to supply the starch works. This type of system is to be found in Thailand, in most Latin American countries and, in a few instances, in Africa. Tractors with appropriate machinery are generally used for field preparation (possibly for burying green manure, but primarily for ploughing and loosening the soil), transporting the cuttings, the first maintenance operation (which usually consists of ridging) and transporting the harvested crop.

The partially mechanized farms generally use improved varieties, fertilizer and sometimes herbicides.

### Wholly mechanized cultivation

Completely mechanized cassava production has not yet been attained on an everyday commercial scale. It is usually the subject of experiments or pilot schemes. The special equipment needed for planting, harvesting and loading is not available everywhere and is still being refined or perfected in a number of countries.

*Permanent farming systems must be developed:*

*(a) where insufficient land is available to permit long fallow periods (which to some extent enabled soil fertility to be regenerated)*

*(b) where relatively high output is needed from a limited area of land (processing industry).*

## 1.3. *The plant and its main characteristics*

### Growth and growth period

There are certain features peculiar to cassava with which one must be familiar if the crop is to be grown in a rational manner.

Cassava is a perennial plant which may continue to grow for a number of years if it is not pulled up. Its stems may grow to a height of 3 to 4 m. In contrast to other crops, instead of seed being sown in order to harvest fruit, it is stem cuttings which are planted, in order to obtain tuberous roots.

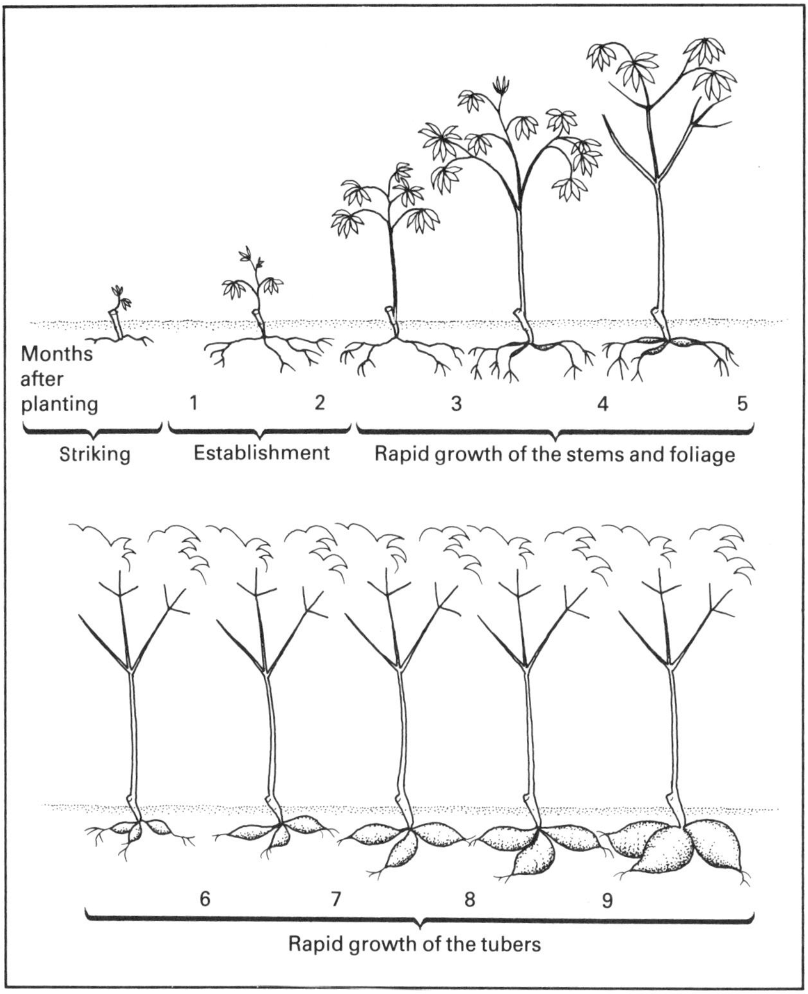


Fig 3 *Development of the plant*

When a cutting is planted, the roots grow first and then the buds which will produce the stems appear. This is the *striking* phase, which takes from three to six days. During the first month, the roots spread out rapidly, at first horizontally, then more or less vertically. During this period the stem grows slowly — it is known as the *establishment phase*, during which the plant lives mainly on the reserves contained in the cutting. The third phase is that of *aerial development*, which lasts for about three to four months after planting. During this phase, the stems grow extremely fast and the plant creates the foliage which will enable it to

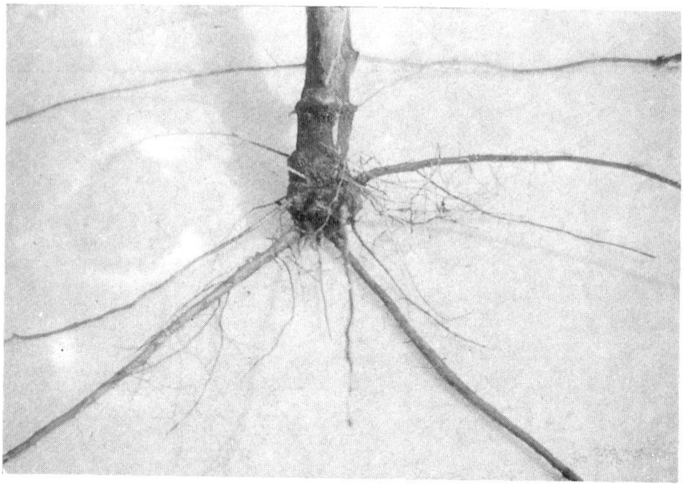

Fig 4 *Rooting at three months*

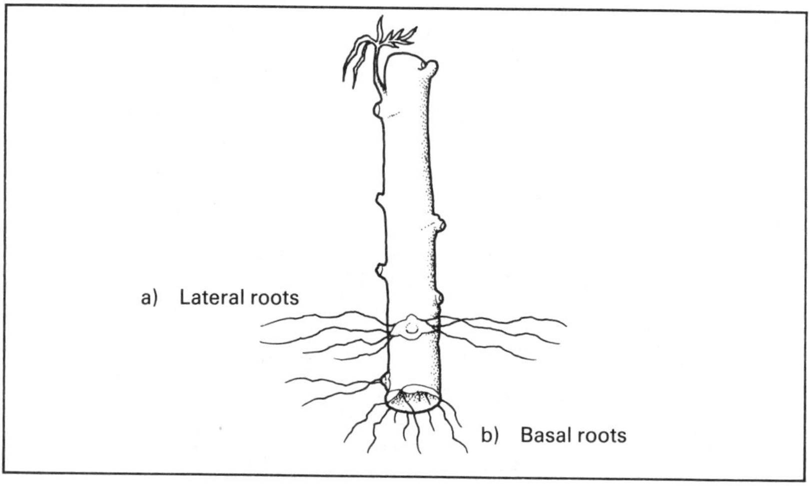


Fig 5 *Cutting rooted 15 days after planting (after Cours, 1951), a) Lateral roots b) Basal roots*

Fig 6 *Tuberous roots*

produce the reserves that it will store in the tubers. The next phase is that of *tuber development*. Some roots start to swell during the preceding phase, but this process accelerates when the foliage is fully developed, i.e., when it completely covers the ground. During this phase, storage of starch in the tubers is irregular, varying with the age of the plant and also according to the season.

Storage is at its peak when the foliage reaches its optimum size. If the branches and leaves are extremely developed, too much of the dry matter produced by the leaves goes into the aerial part. If they are insufficiently developed, not enough dry matter is produced.

Breeders are endeavouring to create plants that have an optimum structure and keep it for as long as possible.

Tubers grow very rapidly around the four to eight month stage, after which they continue to grow, but more slowly. As they get older, they become fibrous.

When a dry season occurs, the tubers stop growing and sometimes decrease in weight, owing to a loss of water, and their starch content increases. When the rains return, the plant produces new branches and leaves, and the dry matter in the tubers sometimes decreases. The tubers do not resume their growth until the foliage has built up again. Thus, growth is generally slower than during the preceding rainy season.

The dry season is a good time for harvesting as the tubers are less wet, richer in starch and have stopped growing.

| |
|---|
| *For proper crop management, it is important to know the rate at which starch is being stored by the plant.* |

Varieties with excessively poor aerial development sometimes develop and adapt better in situations where the average temperature is higher. Conversely, where stem growth is luxuriant and mobilizes too much of the dry matter that is produced, at the expense of the roots, the varieties in question might be more suited to cooler climes.

> *The varieties best suited to a given ecological situation are those which, in that environment, maintain a good balance between the production of new leaves and stems and that of tuberous roots.*

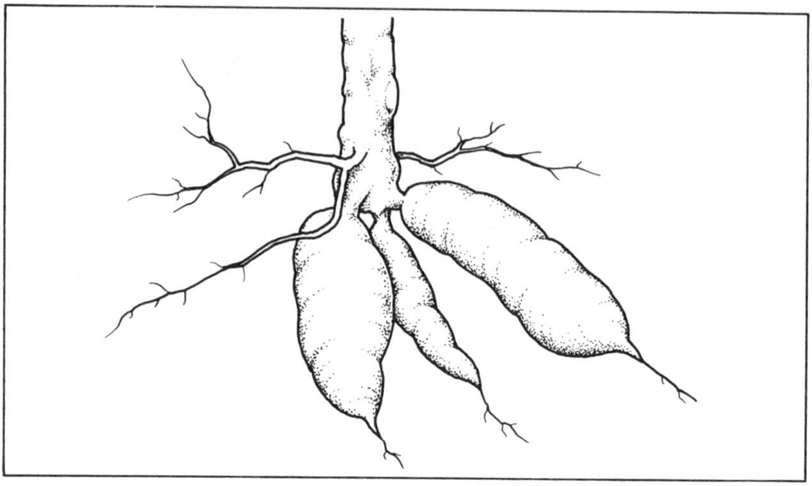

Fig 7 *Tuberous roots*

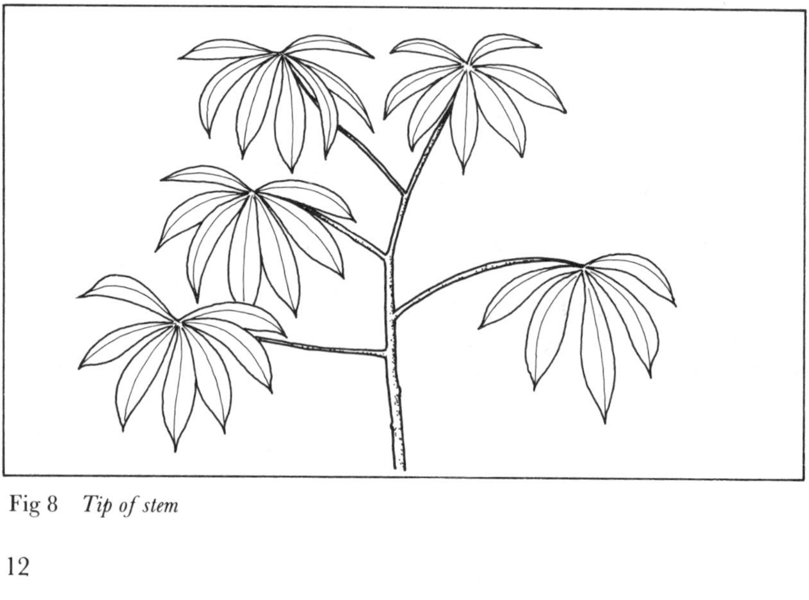

Fig 8 *Tip of stem*

## Drought and excessive water

Cassava withstands drought well because of its deep roots, but very severe droughts may cause partial or total defoliation. Even if that happens, the plant seldom dies, but yields may be seriously reduced. Periods of drought which do not affect the aerial vegetation seem to have little impact on yields, but those which reduce the amount of foliage do slow down the growth of the tubers. The alternation of such periods of drought with periods of recovery is very detrimental to production.

Cassava stands up well to high rainfall, provided that the soil is well drained. However waterlogged soils, can cause the roots to rot.

## Cassava, an exhausting crop

Cassava does exhaust the soil of nutrients, but it is a crop which can use 'worn-out' soils better than others.

Nutrient uptake by cassava is comparable to and sometimes even lower than that of other crops for the same amount of useful product harvested. For instance, 12 t of cassava and 4 t of maize (total fresh matter) are roughly equivalent in terms of useful dry matter. On that basis, uptake by the two crops is as follows:

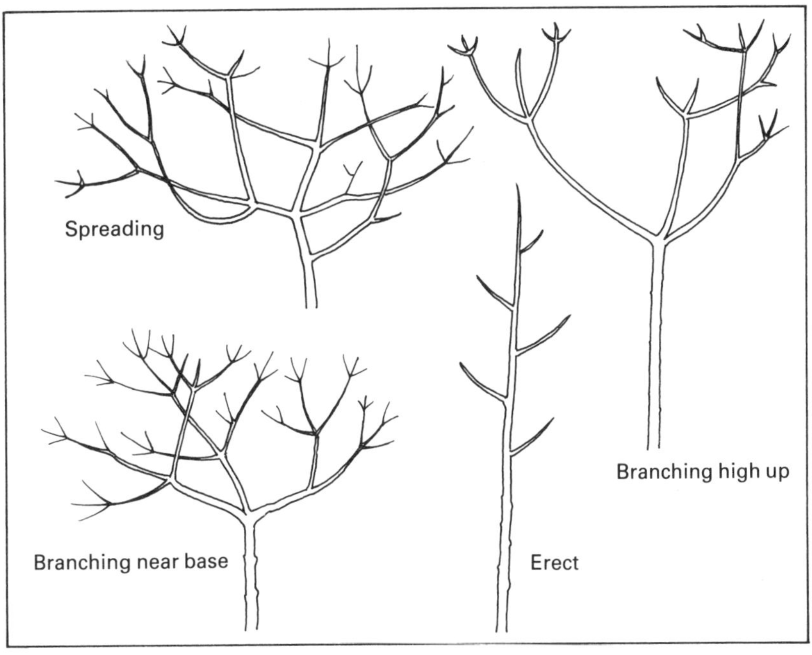


Fig 9 *Different types of branching and habit*

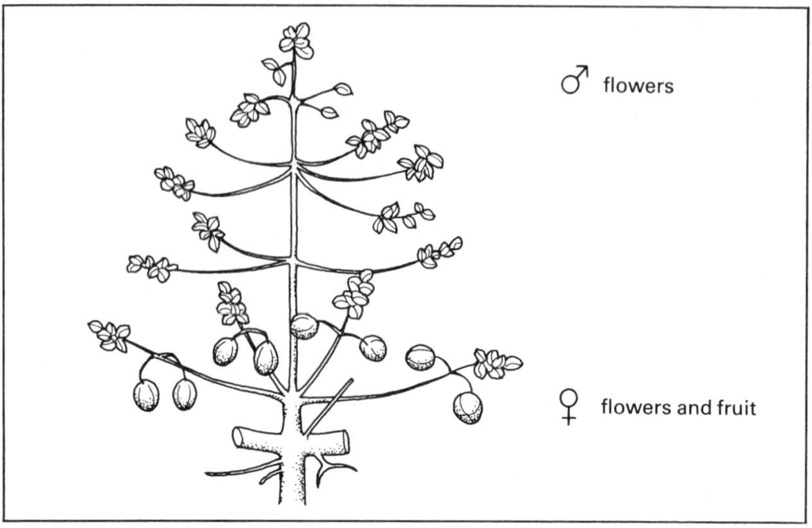


Fig 10 *Inflorescence*

Fig 11 *Cassava fruit*

**Table 2**

| | | cassava | maize |
|---|---|---|---|
| **Nitrogen** | N | 55 kg | 150 kg |
| **Phosphate** | $P_2O_5$ | 30 kg | 55 kg |
| **Potash** | $K_2O$ | 90 kg | 85 kg |
| **Lime** | CaO | 30 kg | 15 kg |

Cassava extracts these quantities of nutrients over a fairly long period (8 to 10 months) and is capable of removing them from fairly poor soils. Under the same conditions, maize must have the same minerals available over a shorter period of time (4 to 5 months), which means that soluble fertilizers that can be rapidly assimilated must be applied.

On average, for a tonne of tubers harvested, cassava removes from the soil approximately:
4.5 kg of nitrogen,
2.5 kg of phosphate,
7.5 kg of potash,
but a large proportion of these elements are to be found in the stems and the leaves. A tonne of tubers contains only about:
2 kg of nitrogen,
1 kg of phosphate,
4 kg of potash.

Nitrogen is useful for growth as it is needed by the plant particularly during the aerial development phase. Phosphorus and potassium are useful for starch formation and tuber growth. Potassium is the most important nutrient, but the plant always tends to remove more than it really needs. When the tubers are growing, a surplus of nitrogen is undesirable because this will encourage aerial growth at the expense of tuber growth.

*Cassava appears to be an exhausting crop, because it responds poorly to the application of fertilizer and does not encourage farmers to return to the soil the nutrients that have been removed.*

## Weeds

With the shifting cultivation systems, there are few weeds immediately after the land has been cleared, but after a few years fast-growing annual grasses start to spread and force the plot to be abandoned. Under a permanent farming system, an effective war must be waged against this invasion by weeds.

Regularly weeding fields throughout the growing season enables the highest yields to be obtained, but the most important weeding is that done when the crop is in the early stages of its growth.

A field that has been weeded only during the first four months generally produces only 10% less than if it had been regularly weeded right through to harvesting. After the first four months, which is when the plant is growing rapidly, cassava is less sensitive to competition from weeds. Moreover, the fact that the leaves are fully developed impedes weed growth. After the fourth month, cassava can look after itself, weeding becomes difficult and is generally no longer worthwhile. However cases do occur when the crop is troubled by late weeds as a result of heavy losses of leaves caused by a dry season, a cold season or parasite attack. Late weeding may then be necessary.

## *1.4. Cassava diseases and pests*

### Diseases

The main diseases that are harmful to cassava are African mosaic disease, bacterial blight or wilt and various types of root rot.

**African Mosaic**

This disease is caused by a virus which is inoculated into the plant by a small white insect of the genus *Bemisia manihotis*. (see Insects page *18*) The severity of the attack depends on:

(a) the susceptibility of the variety in question,
(b) whether the disease is present in the cutting when it is planted,
(c) whether it is inoculated by the *Bemisia* into plants that are already developed and have already started to produce tubers.

The most serious attacks are found among the most sensitive varieties and when the cuttings used already contain the virus. Heavy yield penalties are sustained from this disease in many parts of Africa.

The incidence of the disease may be reduced by choosing the most tolerant varieties and selecting cuttings from plants which do not show any signs of mosaic. Generally speaking, in the traditional cassava-growing regions fairly tolerant varieties exist that have been selected by

Fig 12 *Leaf deformity caused by mosaic virus*

farmers from observation over many years. The research stations can supply varieties which are both tolerant and productive.

**Bacterial blight**

This disease (*Xanthomonas campestris pv, manihotis*), which, as its name suggests, is caused by a bacterium, is currently the most serious disease threatening cassava in Africa.

Infection may result from penetration of the bacteria via the leaves or from the use of contaminated cuttings.

Where the bacteria penetrate into the leaves of a healthy plant, they cause angular spots, which grow and turn brown. As the infection progresses, small gummy balls can be seen on the petioles and the young stems. Finally, when the disease has spread throughout the plant, cankerous lesions appear on the stems, the tips of which dry up and die.

Where the infection comes from the cutting, the young leaves at the tips of the stems suddenly wilt, particularly during the rainy season. Cankers may then appear and the branches blacken and dry out.

An effort may be made to avoid the most serious cases of the disease by using only cuttings from plants which show no sign of bacterial blight. However, the only truly effective method is to use resistant

Fig 13 *Bacterial blight of cassava* (Xanthomonas campestris pv, manihotis), *Angular spots on leaves Blight burns*

varieties where they exist. The research centres are creating such varieties which will generally be resistant to mosaic virus too.

**Tuber rot**

The tubers may rot as a result of being attacked by various species of fungus. Such attacks are fairly rare and occur mainly, depending on the species, on wet poorly-drained land or soils resulting from the clearing of forest; where they occur, they may entail considerable losses. Rot which is not caused by microbes is also occasionally observed (aseptic necrosis).

## Insects

A very large number of insects can live off cassava. Some may cause damage, but they seldom have much effect on yields. Control with insecticides is generally not economic.

In Africa, the main pests are:

(a) white flies (*Bemisia manihotis*) which are the carriers of mosaic virus.

Fig 14 *Aseptic necrosis of the tuber*

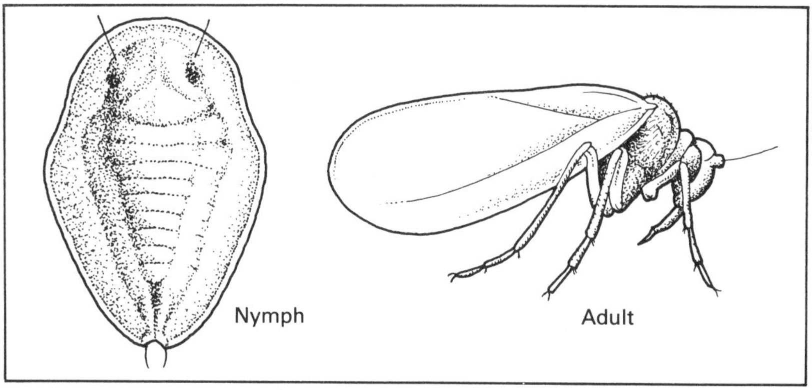


Fig 15 Bemisia manihotis — *adult and its nymph, found on a cassava leaf*

Fig 16 *Mealy bug* (Phenacoccus manihotis), *Young branch attacked by bugs*

(b) termites, which may attack the cuttings, particularly when conditions for striking are poor;

(c) *Acrididae* which can devastate all crops, but are invariably less harmful to cassava because its tubers are beyond their reach;

(d) bugs, particularly:

- a scale insect (*Aonidomytilus albus*), which may cover the stems and leaf-stalks and cause weakening of the plant;
- a mealy bug (*Phenacocus manihoti*), which has recently appeared in Africa; these bugs cover the tips of stems, surrounding themselves with white flakes. The leaves almost completely disappear during the dry season and production is reduced.

These bugs have external coverings which resist the penetration of insecticides.

Cuttings taken from plants infested with these bugs do not strike easily.

The only conceivable way of controlling bugs is to use biological methods consisting of encouraging their natural parasites. Research on this subject is under way, but such methods cannot be used on a farm scale and will have to be organized on a regional basis.

### Mites

Three species of mite originating in America are now to be found in Africa. First reported in East Africa, they are now encountered in other regions. The most important species is *Mononychellus tanajoa*, which is small and green. The mites attack the underside of the leaves, which become deformed as in the case of mosaic. The damage may be considerable (40% loss) in the event of large-scale invasions. Biological control is being studied.

### Other pests

Rodents (rats and agoutis) or various species of wild pigs may ravage the crops.

## Toxins

### Hydrocyanic acid (HCN)

Cassava contains a cyanogenic glucoside, a substance which, as it breaks down, produces poisonous hydrocyanic acid. This substance is present throughout the plant, but the amounts vary from organ to organ. The leaves and the tips of the stems generally contain rather more than the tubers, whose rind contains more than the central part. Glucoside content also differs according to variety.

Generally speaking, the varieties which contain little glucoside are described as sweet and those which contain more as bitter or very bitter. However the correlation between taste and glucoside content is poor.

Certain varieties which appear to be fairly sweet may have average to high glucoside content, simply because they contain slightly more soluble sugars than the others. The preparation or processing undergone by cassava before it is consumed gets rid of most of the toxic constituents (drying of thin chips or 'cossettes', pounding or grating of fresh tubers, followed by squeezing out the liquid and drying, soaking or steeping whole or peeled tubers for a few days, boiling in a large amount of water). With regard to the leaves, which are often consumed, wilting after harvesting and boiling in water also get rid of virtually all the toxic constituents.

*The absorption of 60 mg of hydrocyanic acid, which may be contained in 100 g to 1 kg of fresh tubers, according to variety, may lead to acute poisoning.*

**Preserving the crop**

Tubers that have been harvested remain in good condition for about two days. Then, if they are cut open, bluish dots and lines will be seen first, then the tissue will turn brown and, finally, it will ferment and decompose. Some techniques may extend the keeping time slightly but none is perfect. The tubers must therefore be processed or consumed as soon as possible after harvesting.

# 2 Cultural practice



## 2.1. Introduction

We shall now examine cultural techniques that are appropriate to situations in which the economic environment encourages farmers to increase agricultural productivity and hence modernize and intensify their way of farming. Such a situation is that in which a relatively large proportion of the production is marketed or undergoes industrial processing. The money earned in this way can then be used to purchase further inputs such as fertilizer, pesticides and machinery.

In addition to the large-scale production units that supply industry, it is mainly small and medium-sized farms, whether family units or cooperatives, which will have to employ techniques that are intended to increase productivity in order to provide regular surpluses beyond their own personal consumption needs.

Inevitably this will be some kind of permanent system, because significant increases in productivity are impossible with shifting cultivation, and also because collection, where it involves fresh roots, requires production to be concentrated geographically. In principle, modernized cassava production, where it is carried on in some form of money-based economy, may be undertaken using manual techniques, draught animals or tractor power.

Indeed, intensification of manual cultivation does enable yields to be increased, but, in return, it requires fertilizer and a greater labour input, so that from the point of view of economic development and improving income levels, some degree of mechanization is necessary. We shall therefore be considering principally mechanized farming systems, particularly those using tractor power, as in Africa cassava can seldom be grown using draught animals, particularly in the tse-tse fly zones.

However, the emphasis will be placed mainly on the results to be obtained with different cultural practices rather than on specific types of equipment to be used, so that the recommendations that are made apply as widely as possible.

## 2.2. Choice of region and soil

*Intensive production requires favourable climatic and soil conditions. However, compared with other plants, cassava exhibits considerable tolerance of these factors.*

### Climate

The amount of annual rainfall needed to grow a good cassava crop varies from 1000 mm to more than 3000 mm.

The distribution of the rainfall is an important factor affecting productivity but it is impossible to lay down hard-and-fast rules. For example, one should not necessarily avoid situations where the dry season lasts longer than a certain number of months. Poor rainfall distribution may be offset by the soil characteristics (topography or texture). Cassava stands up well to a prolonged dry season, in as much as this seldom causes the plants to die. However, periods of drought which

(a) bring growth to a standstill
(b) cause leaves to be lost and then
(c) growth to be resumed, (particularly if these phenomena occur repeatedly),

are very detrimental to yields, even in situations where the overall amount of rain is considerable. Where precipitation is low (1000 mm), it must be well distributed. The ideal conditions are rain every two to three days, with large amounts of sunshine in between, without a marked dry season or with a dry season lasting no longer than two months.

The harmful effects of repeated periods of drought may be offset or attenuated by ground or soil conditions:

- topography: if the land is situated at the foot of a slope, where the groundwater provides a regular supply of water, but without leading to surface waterlogging;
- texture: if the soils are deep and clayey, with good water reserves.

Where the annual rainfall is very high, the texture and topography of the soils must be such that drainage is good.

The risks occur where:

(a) the rain is evenly distributed, but there is too much cloud and the amount of sunshine is inadequate;
(b) the rain is unevenly distributed and occurs in violent downpours leading to run-off and erosion or waterlogging of the soil, all of which have an adverse effect on yields;
(c) rainfall is such that it seriously hampers mechanical operations, particularly field preparation, harvesting or transporting the crop.

The most favourable conditions often occur where the climate is of the Guinean type, where total precipitation does not exceed 1500 to 2000 mm a year.

### Soils

Of course deep soils with good structure and a balanced texture, which are chemically rich and well-supplied with organic matter, are ideal. However, cassava is a hardy plant and often the best soils are reserved for more demanding and more profitable crops.

In Latin America, cassava is being increasingly considered for making the most of very acid land, which would otherwise remain unused.

Indeed, cassava tolerates a wide range of soils, from slightly alkaline to highly acidic, from clayey to fairly sandy, but it will not tolerate poorly drained soils.

In regions where precipitation is irregular, one should chose deep soils, without an impermeable layer close to the surface, with a good waterholding capacity and hence fairly clayey. Where precipitation is plentiful, lighter, well-draining soils should be chosen. However, extremely clayey soils, particularly if the structure is poor, are unsuitable because of the mechanical difficulties involved at harvest-time.

### Sloping ground

Cassava offers the soil little protection against erosion. The decision as to whether or not to cultivate slopes with a gradient from 3.5 to 10 percent depends on numerous factors, which must be assessed according to local conditions, such as soil structure and susceptibility to erosion, nature of the precipitation, possibility of timing planting so that the crop is well advanced when the most erosive rains occur and, lastly, the possibility of using anti-erosive cultural techniques (ploughing along the contour line, ridging, etc.) or carrying out other measures to combat erosion.

One good anti-erosion measure consists, in production systems where the cassava alternates with fallow or a fodder crop, of carrying out this rotation on adjacent strips along the contour line, the steeper the slope and the greater the risks of erosion, the narrower the strips.

There are numerous instances of severe erosion resulting from Cassava on steep slopes in the Caribbean area.

## *2.3. Timing of operations, cropping plan, rotations*

In a given situation, the farmer or agricultural adviser who is establishing a cassava plantation or wishes to improve an existing plantation must decide on his cropping plan and the rotations to be adopted.

The questions which need to be asked are as follows:

(a) When is the farm to supply its products?

(b) After how long a growth period are the crops to be harvested?

(c) Is there, in the region in question, any particular time which is most favourable for planting cuttings?

(d) What other lines of production or what cover crops are possible, and where do they fit into the rotation or farm plan?

## Timing and staggering of harvesting

Cassava, whose tubers do not keep after they have been lifted, must be harvested when it is going to be used. On the family-run smallholdings, where cassava is intended principally for personal consumption, the cassava is harvested regularly throughout the year for direct consumption and for making products which can be prepared whatever the season (gari, cassava balls or sticks). Harvesting takes place during the dry season, however, where the products need to be dried in the sun (cossettes, fufu, etc.). Farms which are geared to meeting market requirements (those considered here) may be of two main types:

(a) those where the harvesting is concentrated in the dry season,

(b) those where the harvesting is spread throughout most of the year.

The first type of operation corresponds to situations where the tubers are intended to be dried in the sun, either by the farm or by a cooperative organization, or even by the purchaser. However, often an effort is made to spread out harvesting as much as possible, in order to make the best use of the manpower and machinery available, even if shelters have to be provided for drying the cassava during rainy periods. This is the practice in Thailand.

The second type of operation corresponds to situations where there is a permanent demand for fresh tubers, particularly where an industry (e.g., starch manufacturers) is to be supplied. In this latter case, it is the industry which requires production to be spread out over as long a period as possible, because the longer their plants are working, the more quickly they pay for themselves. However, this also suits mechanized farms, which do not need so much machinery if it can be used over a longer period of time.

## Growth period, age of plants at harvesting

We have seen that once the aerial part of the plants is well developed, the tuberous roots start to grow rapidly and continue to do so for a number of months, depending on environmental conditions and variety, then growth slows down but without stopping altogether. Finally, the roots become increasingly fibrous and may be liable to rot.

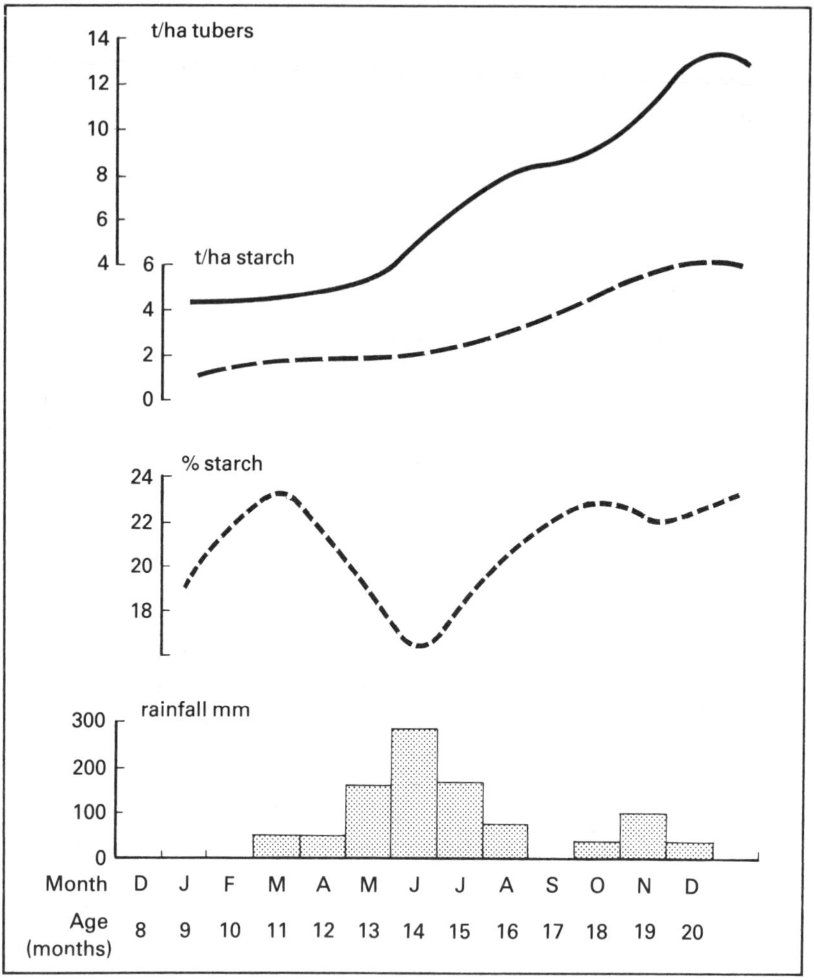


Fig 17 *Effect of climate on the production of tubers and starch*

Thus, there is no hard-and-fast growth period after which the crop must be harvested.

Short growth periods, for instance 8 to 10 months, result in poor yields. If there is no particular reason to harvest at that age, it is extremely desirable to wait as the crop requires virtually no maintenance, the roots are developing and yield increases are obtained at no cost to the farmer. From 15 to 20 months onwards, depending on variety, the increases in output to be expected from extending the cycle may be insufficient to justify the fact that the land cannot meanwhile be used for other purposes. The growth period, or rather the range of growth periods adopted on a given farm, will fall somewhere between these two extremes.

## Time of planting

Where the climate is normally wet and there are no marked dry seasons, cuttings may be planted virtually all the year round. Yields, for plants of the same age, will be much the same irrespective of when the cuttings are planted. However such conditions are rare. Usually, there is one dry season, and sometimes there are two.

Where there is a dry season, the first rainy period that follows is often the best time for planting cuttings in order to obtain a high yield. For plants that are the same age at harvesting, the cuttings planted later produce lower yields than those planted earlier. The later they are, the lower the yields. However it is usually impossible to plant all the cuttings at the most opportune time, as farmers are limited by the availability of labour or the capacity of the machines.

The cuttings are usually planted over a period of one to two months. In some cases, it may be desirable to plant some of the cuttings at the optimum time and the others later on. The cuttings planted later will be left in the ground longer than those planted earlier, in order to make up for the consequent loss of yield.

Lastly, in some cases there are no marked differences in yield between crops from cuttings planted early and those from cuttings planted late. Usually, particularly where a large farm is to be organized, the effect of time of planting on yields must be verified experimentally.

## Crop cycles, timetables

The farmer is generally obliged to grow, at the same time, crops that will be following a different timetable, and whose cropping cycles may also be different.

In the extreme case where cuttings are planted over a single period of two months and production is to be spread throughout the year, there would be a difference of ten months between the longest and shortest growth periods. Where the cuttings are planted over two separate periods, with an interval of several months in between, the differences in the cropping cycle may not be so great.

Any cassava enterprise will consist of crops that, although involving the same species, generally exhibit significant differences. Operations such as field preparation, planting, maintenance and harvesting may require different techniques if they occur at different times of the year.

## Cropping systems

Economic reasons may lead to cassava being grown either by itself as a monoculture or alongside other livestock or crop production enterprises.

It is often in the interest of large-scale mechanized farms that use paid labour and supply a processing industry to produce cassava alone, in order to simplify management, machinery requirements and maintenance. In some cases (industry using by-products to produce animal feed), it may prove worthwhile to grow maize as well, from the point of view of both the farm and the factory.

The growing of cassava should be combined with other enterprises if part of the production is for personal consumption, but also in some cases for purely technical reasons (soil conservation).

It has often been found that if cassava is grown continuously on the same ground, after a number of years yields will steadily decline. Where depletion of the soil's chemical reserves is not responsible, it is not always easy to explain why this decline in yields occurs. It may be a matter of physical deterioration, as the soil is very exposed to erosion at the start of the cycle and after harvesting. The reduction in organic matter content has often also been blamed, as cassava generally responds favourably to applications of organic matter.

In view of these observations, it is recommended that cassava should not be grown continuously.

The simplest solution is to lay the land fallow for a year or two after two or three cassava crops. However, there are drawbacks to fallowing: it is not a good way of dealing with erosion and it makes a mess of the land and makes it more difficult to control weeds. Instead of fallowing, a crop may be grown for digging in (green manuring). Depending on the ecological conditions, the green manure may be planted for only a few months, this being repeated between each crop. Alternatively, there may be a longer regenerative break lasting for, say, two years, after every two or three cassava crops.

Long regenerative breaks are probably more effective for conserving or improving fertility. Consideration may also be given to making the most of them by using them as fodder, leading to mixed cassava/livestock enterprises.

The introduction into the rotation of artificial grassland grazed by live-stock is surely one of the most effective techniques for improving fertility. Thus some people have considered combining cassava-growing with the rearing of cattle or sheep. Such solutions do, however, pose technical problems — notably with regard to enclosures and watering facilities, which need to be temporary — that are not always easy to resolve. What tends to happen is that the grass is harvested and the animals fed outside the plots. This method prevents the animal waste from being deposited directly on the plots. It is acceptable provided that all the dung is spread on the fields. Fodder crops remove from the soil quantities of mineral matter which are generally far higher than the amounts removed by cassava, so it is essential to return them to the soil if fertility is not to be rapidly reduced. Cassava may also be 'alley cropped' with other species across contours.

The main species that are suitable for rotating with cassava are as follows:

**food crops**: maize, rain-fed rice, cowpeas (*Vigna unguiculata*), soya beans, groundnuts;

**short-term green manures**: normally legumes are chosen, such as *Vigna*, *Crotalaria*, hyacinth beans and fodder soya;

**long-term green manures**, such as *Pueraria*, *Centrosema* and in particular *Stylosanthes guyanensis*;

**fodder species**: the species referred to above may be used, particularly *Stylosanthes*, but also *Gramineae* such as *Brachiaria* or various *Panicum maximum* ecotypes.

## Cropping plans, rotations and patterns

*The cropping plan must be studied carefully and then followed rigorously.*

If production is to be regular, at set times of the year, and certain technical constraints are to be respected — such as not having to harvest cassava which is too young or too old, to plant cuttings at the most favourable times and to rotate the cassava with another crop, a fallow or a green manure — the rotation which will enable one's objectives to be achieved must be defined precisely.

Working out a cropping plan and the necessary rotation is more complicated with cassava than with an annual crop which has a constant growth period of less than a year and occupies a fixed place in the farming year.

Often, it is because insufficient attention has been given to this matter that farms, sometimes those run as pilot schemes, find themselves with no crop one year and twice the expected crop another year.

The appendix gives a number of examples of rotations and patterns corresponding to some of the situations encountered, which differ according to whether all-the-year round production is required or whether harvesting can be limited to one specific season, whether or not there are periods when the cassava must be planted and how other crops, fallowing or green manures fit into the rotation.

These examples also provide an opportunity to show how to work out a coherent rotation using charts.

## 2.4. *Selection of planting material*

Productivity depends on the cultural conditions but also on the inherent suitability of the planting material used, which is related to its genetic make-up. The genetic factors of productivity are mainly the ability to produce dry matter, adaptation to the physical environment and tolerance of or resistance to disease. Foliage growth — not too luxuriant, nor too sparse — may provide a good indication of whether the plant is suited to its environment.

The main criterion for choosing a variety is of course productivity in terms of dry matter or starch, i.e., of useful product. A second criterion may be earliness, for in a given ecological situation, some varieties are earlier than others.

It can be seen from the above graph that for early harvesting (between 8 and 14 months), variety A is superior to variety B, but for later harvesting variety B is superior to A. Varieties may exist which combine both qualities (e.g., type C).

In all the major cassava-producing regions of the world, there is usually a very wide range of traditional varieties which are the result of empirical selection by the farmers from among plants produced from seed growing in fields that have been abandoned. These varieties have generally been chosen for specific qualities which may be technological, agronomic or relate to taste. They are often well adapted to the physical environment. Some are fairly productive; sometimes they are quite tolerant of mosaic but far less often of bacterial blight.

Finally, all the research centres and stations have programmes aimed at creating highly productive, disease-resistant material and selecting

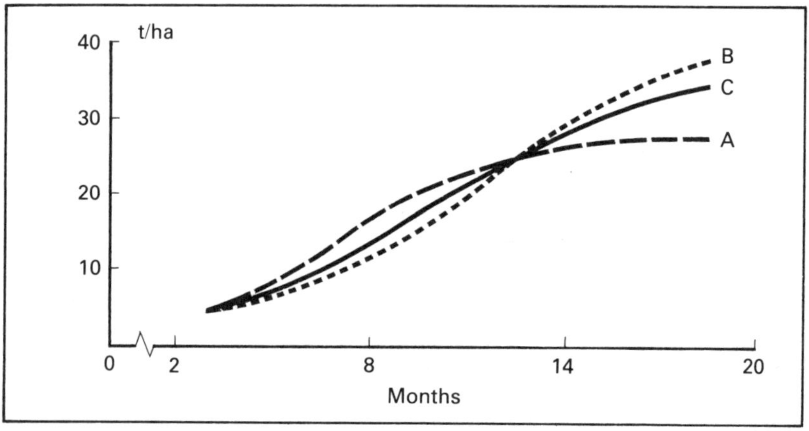


Fig 18 *Yield from varieties with different degrees of earliness*

varieties that are adapted to the ecological situation with which they are concerned.

To find out which is the best variety to grow, contact the research centres or stations. If there is no selected variety, it will be advantageous to use the best local variety. In many cases, it may even prove equivalent or superior to a variety that has been selected in a different ecological situation.

The principal varieties recommended in Africa are as follows:

*Ivory Coast*:

CB: a clone originating in Zaire, and studied for a long time in Ivory Coast.

H57-H58: clones originating in Madagascar.

*Nigeria*:

37 065 – 44 086 – 53 101 – 60 444 – 60 506 – 60 447.

*Madagascar*:

H 54 – H 57 – H 60: clones adapted to altitudes of around 800 m and mosaic-tolerant.

For new cassava varieties see Appendix, page 76.

## 2.5. *Field preparation*

### Objectives

Field preparation covers all the operations involved between the preceding crop and the planting of the cuttings. We shall not be dealing here with the development of uncultivated land; such land may be cleared using a wide variety of techniques, depending on the nature of the original vegetation, the climate and the soil, but such operations are not specific to the crops for which the land is being prepared.

Field preparation has many aims:

(a) to clean the soil by burying the natural vegetation, residues of the preceding crop or a green manure;
(b) to bury organic matter;
(c) To bury artificial fertilizer;
(d) to restore to the soil the minerals contained in the crop residues or green manure;
(e) to combat weeds;
(f) to improve the movement of water through the soil;
(g) to create surface conditions that are conducive to the germination of the cuttings;
(h) to loosen the soil, which encourages tuberization of the roots;
(i) where appropriate, to create ridges.

These aims are achieved, to a greater or lesser extent, by a succession of operations which, chronologically, are as follows:

- operations carried out prior to tilling the soil, the purpose of which is to make the soil workable;
- tilling of the soil proper, the principal function of which is to loosen the soil to a greater depth and combat weeds, but which may also lead to the creation of ridges;
- final cultivation or tilth production, which may complete the destruction of weeds but the main role of which is to break up the surface clods, so that the cuttings strike more easily: ridging may be one of the final cultivations.

## Preliminary operations

The residues left on the soil by an annual crop (maize, soya beans, etc.) may be directly buried by ploughing them in. This does not apply if the preceding crop was cassava or a green manure which has been left to grow for some time. The mass of vegetation may be such that ploughing is impossible. One solution is to destroy it by burning, which is not always possible and often leads to considerable losses of organic matter and nitrogen.

### After a cassava crop

It is recommended that the wood and leaves left over from the preceding crop should be buried, as they represent an organic contribution to the soil of 15 to 30 t/ha of green matter and a substantial amount of minerals (30 to 60 kg of N, 15 to 30 kg/ha of $P_2O_5$ and 35 to 70 kg/ha of $K_2O$.).

The vegetable mass left after harvesting cassava must be chopped up not only to facilitate burial but also to prevent fragments of stem from propagating. The aerial parts must be chopped up, preferably prior to harvesting, particularly where this is done mechanically. Where the crop is lifted by hand, chopping after harvesting may offer the advantage of leaving the debris on the surface of the soil, thus facilitating drying prior to burial and reducing the risk of fresh growth.

The equipment used to chop up the aerial vegetation of the cassava may be of the rotary cutter (vertical axis) or rotary chopper (horizontal axis) type.

Generally speaking, the rotary cutters work quickly, have only a moderate power requirement, but do not shred the stems sufficiently to prevent them from propagating. The rotary choppers need more power, particularly if the vegetation is dense, but the job they do enables the tubers to be lifted and the vegetation to be buried under optimum conditions.

These machines are often mounted behind the tractor, which means that it has to force its way through the vegetation, with the risk that the

Fig 19 *Young cassava plantation in Zaire*

Fig 20 *Harvesting cassava by hand on a state farm in Zaire; preliminary phase; cutting down the aerial parts*

engine will not be cooled properly. Also the stems will be crushed by the front of the tractor and because they are lying flat, they will remain unscathed by the machinery.

It is recommended that the rotary choppers be mounted in front of the tractors or, if mounted behind, be offset to one side. With a rotary chopper that has a 1.50 m cutting width it takes about three hours to do one hectare.

## After a green manure

Depending on the type of green manure used and the extent of its growth, burial may be carried out directly or after cutting back the vegetation.

In the case of a two-and-a-half to three year old *Stylosanthes* which has not been used for grazing, the plant mass is so great that direct incorporation is impossible. The volume of this mass may be reduced by using a rotary chopper or cutter, but more simply by making a pass with a machine (such as a ring roller, a roller chopper or even a disc harrow) that is capable of breaking up or crushing the stems, which will wilt the following day.

## Tilling the soil

The soil is tilled by means of ploughing (loosening of the soil with inversion of the top layer) or by means of tine or disc cultivations (loosening of the soil without inversion). Ridging is an intermediate form of tilling in terms of inversion, but it is aimed at altering the relief of the land in a particular way. A ridge consists of a continuous raised strip between two furrows. The ridges are where the rows of plants grow, the furrows are the spaces between the rows.

## Ploughing

Inversion of the soil is necessary in order to bury the residues of the preceding crop, a green manure, dung or artificial fertilizer, and it loosens the soil and in particular controls weeds better than tine cultivations.

Good ploughing should achieve unbroken set-up furrow slices, in other words the inverted strip of earth must be right up tight against the preceding one, without leaving any gap. Ploughing depth is also an important quality criterion, since it determines the volume of soil loosened. For cassava, the recommended ploughing depth is 20 to 30 cm.

## Tine or disc cultivations

Pseudo-ploughing may be preferred to ploughing proper if the ground is clean, if it can save time, and if the final operation complements the loosening of the soil and consists in particular of ridging.

## Ridging

In a number of countries, cassava is traditionally grown on ridges because, if the work is done by hand, it is an easier way of loosening that part of the soil where rooting takes place. Cassava yields are not normally any higher where the crop is grown on ridges rather than on the flat, except perhaps where there is a danger of surface waterlogging. Where cultivation is partly mechanized, if the cuttings have to be planted by hand, ridging offers the advantage of marking out the rows for planting. For this reason, it is often recommended where a cooperative or community organization prepares mechanically land that is intended to be cultivated manually by villagers, as is currently the case in Ivory Coast.

Ridging may be carried out directly or following ploughing or pseudo-ploughing.

**Implements for tilling the soil**

Ploughing is done with either mouldboard or disc ploughs. Mouldboard ploughs are preferred where the soil structure is delicate, while disc ploughs are more suitable where stumps have not been grubbed up properly from the soil.

Cultivation may be carried out with disc harrows or rigid-tined implements.

## Breaking down the soil

The tilth does not have to be as fine for cassava cuttings as for seedbed preparation. Where the ploughing is done under optimum conditions

Fig 21 *Mechanical earthing-up of cassava*

of soil moisture content, according to its physical characteristics, on relatively weed-free ground shortly before the cuttings are to be planted, it is even possible to do away completely with further cultivations. However, in a number of cases — where the ploughing was not done properly (clods too large), where weeds have grown up again or where mechanical planters are used which require an even surface — the ploughed land must be worked down. In some cases, ridging may be used as a technique for further breaking down the soil (planting of cuttings by hand), in which case it is done immediately after the soil has been tilled.

### Implements for working down ploughed land and producing a tilth

There are many implements which may be used, either with discs or tines, but in both categories they may be heavy or light. Thus it is the light disc harrows or various other types of harrows which are strictly speaking the tilth-producing implements.

The choice of equipment for working down ploughed land to some extent depends on the ploughing implements that are available but also on the amount of weeds and the soil's susceptibility to erosion. Generally speaking, tined implements are preferable for eliminating *Gramineae* with rhizomes and are less likely to contribute to erosion than implements with discs.

## Timing field preparation

The timing of field preparation is dictated by the cropping plan. A few examples will indicate what the consequences are:

**1st case**: harvesting is to be spread throughout the year and cuttings can be planted all the year round.
The preparatory work for each of the different fields takes place at different times. The various operations of tilling and breaking down the soil can follow one another in rapid succession and be done over a short period of time. It is, however, possible that certain times of the year will be unsuitable for ploughing, for instance because of abundant erosive rainfall. Growers may then have to modify the preparatory techniques at that time of the year or even modify the programme for planting the cuttings.

**2nd case**: harvesting is to be spread throughout the year and the cuttings are planted at two stages — after the dry season and in the middle of the wet season.

This is the most common case. Let us also assume that the rotation consists of two cassava crops and one green manure.

In this particular case, the preparatory work will be done under conditions which may differ considerably. The following six situations may be distinguished:

(a) preparation during the rainy season, after a green manure crop;
(b) preparation during the dry season, after a green manure crop;
(c) preparation during the rainy season, after cassava harvested a number of months previously;
(d) preparation during the rainy season, after cassava harvested recently;
(e) preparation during the dry season or at the start of the rains, of cassava harvested before the end of the preceding rainy season;
(f) preparation during the dry season or at the start of the rains, of cassava harvested during the dry season.

In particular, the weed problem will be very different in cases (c) and (e) compared with cases (d) and (f). In cases (c) and (e) in addition to the usually perennial weeds existing in the cassava fields at the end of the growth period, there will also be the fast-growing annual weeds that will have sprung up with the rain following harvesting. Cases (c) and (e) and incidentally also cases (a) and (b) represent situations in which the preparatory work can to some extent be staggered. For instance, it may be decided to plough early and delay secondary cultivations until just a few days before the cuttings are planted. This approach is advisable in order to obtain better control of annual weeds and a better breakdown of the organic matter resulting from the residues of the preceding crop or the green manure. Ploughing immediately prior to or at the beginning of the dry season is highly recommended for its favourable effect on soil structure and for conserving moisture deeper down in the soil.

In cases (d) and (f) the timing of the preparatory work is stricter, but the cultivations may be simplified. If the previously harvested fields are fairly 'clean', ploughing can be replaced by tine or disc cultivations.

Thus, on a particular farm, quite apart from the soil conditions, which may vary from one point to another, the conditions governing the preparatory work may often, depending on the cropping plan, vary significantly according to season and, consequently, from one plot to another.

*Keep a careful watch on field preparation techniques and adapt them to the specific soil and seasonal conditions. Secondary cultivations are often unnecessary.*

## 2.6. *Use of fertilizer*

### Which fertilizers should be applied?

In a permanent farming system, fertility can be maintained only by restoring to the soil the minerals removed by the crops.

In the case of rich soils, however, particularly potassium-rich soils, it is not necessary to supply this mineral during the early years, or rather it is not necessary to supply the same quantities as are removed by the cassava.

Similarly, if a leguminous green manure is regularly included in the rotation, nitrogen applications must be lower than the amounts removed by the cassava.

Where the soil is moderately rich and legumes are ploughed in, it is recommended that 10 kg of 10–10–20 fertilizer be applied for each tonne of tubers, which means that if the crop is expected to yield 25 t/ha, 250 kg/ha of fertilizer should be applied, which will supply 25 kg/ha of N, 25 kg/ha of $P_2O_5$ and 50 kg/ha of $K_2O$.

The potash content of the formulation will be increased if production is found to be low despite satisfactory vegetative growth, or the nitrogen content will be increased if vegetative development is poor during the early months.

### When should fertilizer be applied?

Ideally, nitrogen and phosphorus should be applied when the cuttings are planted and the potash four to five months later. However, at that time, which corresponds to the end of major vegetative growth, it is difficult to apply fertilizer by hand and impossible to apply it mechanically.

It is therefore recommended that fertilizer be applied when the land is being prepared, between ploughing and secondary cultivations.

### Method of application

Placement close to the cuttings can be recommended in the case of very low doses of fertilizer intended to correct a deficit in what is still partly a shifting cultivation system.

In a permanent system, where artificial fertilizers are used to conserve the system, it is advisable for the fertilizer to be spread evenly.

It does not really matter which method of distribution is used. Where cultivation is mechanized, the centrifugal distributors are perfectly suitable.

*The use of artificial fertilizer with cassava is a real problem as there is often no immediate return.*

## 2.7. *Planting*

### Source of the cuttings

Cuttings are normally produced by the farm itself. They are taken shortly before planting, from the stems of cassava that is being or is about to be harvested.

Ten good-quality cuttings can normally be taken from one cassava plant, which therefore means that one hectare of the crop should provide sufficient cuttings to be able to plant ten hectares.

A farm on which harvesting is spread throughout the year, and the planting of cuttings is concentrated over a period of two months (which can be regarded as an extreme case) will produce sufficient cuttings (one hectare harvested) for six to be planted.

However, in certain cases the fields being harvested fail to produce sufficient cuttings.

*(a) The fields being harvested are not large enough.*
This is the case where the choice of plants from which cuttings are to be taken is severely restricted. The selection of such plants may be dictated by the desire to take cuttings only from plants which are free from symptoms of mosaic virus or bacterial blight or, alternatively, from a

Fig 22 *Transporting cuttings in Togo*

single variety where several have been mixed together.

Where a relatively large holding is being established, the grower may not have sufficient cuttings of the best variety available or may not yet know which is the best variety and therefore use cuttings from neighbouring farmers.

The range of choice may be extended by employing cuttings taken from plots harvested earlier or not yet harvested.

Note that cuttings may be kept for about three months, provided that whole stems are used, tied together in bundles and stored in the shade, in a cool, moderately well-ventilated place. In exceptional circumstances, the cuttings may keep for as long as five months.

Cuttings may also be taken from plots due to be harvested two months later, without significantly reducing yields.

*(b) Cuttings do not come from the cassava fields.*

In exceptional circumstances, a farm may be obliged to cultivate plots specifically intended for the production of cuttings. It may do so in order to multiply a new variety slightly more quickly than would be possible in the fields where the cassava is grown as a crop. The measures that may be taken towards that end are as follows: prepare a finer tilth than in the case of a normal crop, use short cuttings (15 cm instead of 25 to 30 cm); in order to increase the number, plant at a high density (0.75 × 0.75 m or 0.80 × 0.80 m) and harvest at six to seven months.

Generally speaking, it will be preferable to enlist the services of an agronomic station or a propagation centre, which will be able to use a proper rapid propagation technique.

Fig 23 *Cassava planter. Labourers working, sitting in comfortable seats and shielded by a sunshade. Note the discs for covering the cuttings with earth*

## Rapid propagation technique

This method is unsuitable for a medium-sized farm but could be applied on a large farm which has its own agricultural experimentation unit. The technique described below is the one developed at the International Centre for Tropical Agriculture (CIAT) in Colombia, but there are other techniques which differ only in minor respects.

A hotbed occupying an area of several square metres is needed which can be covered with a plastic film and kept at high atmospheric humidity. The soil of the hotbed must be worked to a very fine tilth and disinfected.

Cuttings, likewise disinfected, with two nodes, are planted in the hotbed at a high density.

Frequent watering keeps the soil moist, but not excessively.

The young shoots produced by the cuttings are cut with a disinfected blade when they are 8 cm long and placed in glass tubes containing sterilized water, where they send out roots. They are then transplanted in a field, at a fairly high density (80 cm × 80 cm), where they will be maintained very carefully, particularly at the time of striking.

The two-node cuttings in the hotbed can produce up to eight successive shoots in four months.

A normal plant can provide 150 two-node cuttings, each producing eight shoots in four months and eight normal plants in a year. A normal plant can produce, under the excellent conditions of this technique, up to 30 normal cuttings. Thus, the propagation coefficient is 36 000 (150 × 8 × 30).

It can be seen that, even with a more modest success rate, this method

Fig 24 *Cassava harvesting with a Russian potato-lifter, mounted behind a 4650M (Zaire)*

will enable 200 to 300 ha to be planted with the same clone from a single plant after a period of about a year.

## Selecting the cuttings

Wood which is too thin and not lignified, from the tip of the stems, and wood from the basal extremity, if the plants are very old, should be discarded.

Large cuttings strike better than small ones, and also produce plants that are more vigorous initially and often higher-yielding, but there is an economic size, which is of the order of 20 to 30 cm.

The best cuttings come from parts of the stem with a diameter of at least twice the diameter of its medulla; they must have at least five buds and be taken from plants without any signs of mosaic virus or bacterial blight.

## Preparing the cuttings

The cuttings are normally prepared entirely by hand.

In the field, in areas free from mosaic and bacterial blight, well-formed stems are cut and the superfluous branches removed. These stems are transported to a sheltered spot or directly to the place of planting. The cuttings proper are prepared only a short while prior to planting. This operation is sometimes mechanized by using circular or blade saws. Whatever is used to take the cutting must make a clean cut, without crushing the stem.

Fig 25 *Preparing cuttings manually*

## Spacing

The optimum spacing is between 0.8 m x 1 m and 1 m x 1 m. The higher planting density will be used for plots that are to be harvested early (within 12 months or less) and the 1 m x 1 m spacing for plots to be harvested later (after more than 12 months).

The same planting densities (i.e., 12500 to 10000 plants/ha) can be obtained with double rows spaced alternately 0.80 m and 1.20 m. Some harvesters are designed for this type of spacing.

## Number of cuttings to be planted together

Cuttings must be planted singly. If several cuttings are planted together, this has numerous disadvantages: it increases the likelihood of there being a number of stems, which is not good for the plant, it means that many more cuttings must be made available and therefore limits the scope for picking out the best ones, and it increases the risk of there being a cutting infected with mosaic or bacterial blight at each planting location.

## Method of planting the cuttings

The cuttings may be placed:

(a) lying down flat, covered over with 8 to 10 cm of earth;

(b) vertically or at an angle, four-fifths of their length being buried.

There are no substantial differences between these methods of planting.

Fig 26 *Traditional farming in Togo; intercropping of maize with cassava*

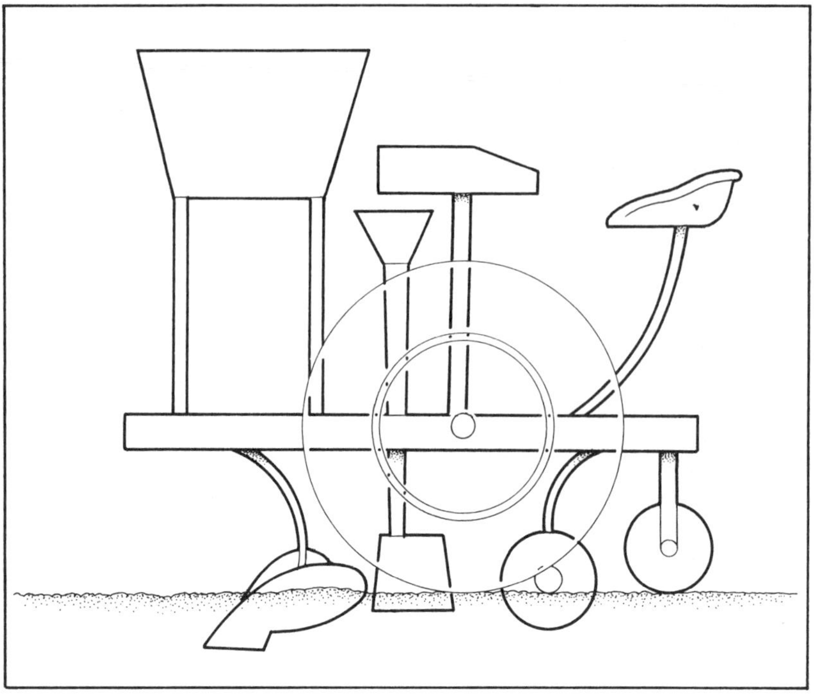

Fig 27 *Diagram of cassava planter (some can plant several rows at the same time). The distributor is fed manually. The machine cuts a furrow, places the cuttings, earths the furrow over with discs and firms the soil with a roller*

Cuttings that are planted flat produce smaller tubers than upright cuttings, but there are more of them and they are less bunched together.

Cuttings that are completely buried are less liable to dry out if there is a risk of drought at planting time. If cuttings are not completely buried, it is possible to check up on the work that has been done in cases where paid labour is being used. The mechanical planters currently in use generally place the cuttings in a horizontal position.

## Planting equipment

Planting is often done by hand, even on mechanized farms. The rows for the cuttings may be marked out with a drill or, as referred to earlier, by means of ridging. A tractor and trailer distribute the cuttings to the fields, where a team of workers plant the cuttings by hand.

The existing planting machines are for planting several rows at the same time, normally three or four.

A three-row machine consists of: a receptacle which holds the cuttings and may be divided into three compartments, and for each row: a seat, a distributor drum driven by the wheels, a delivery chute in front of which is a ploughshare for cutting the furrow and, at the rear, one or two discs for closing the furrow and a heavily weighted roller. Three operators are needed to take the cuttings from the receptacle and feed the drums.

### Striking and shooting

The cuttings send out shoots six to ten days after planting. If they have been properly selected, more than 90 % of the cuttings will normally strike. With this success rate, there is no need to replace any plants that fail.

Normally, it is seldom worthwhile to replace an evenly distributed patch of cuttings that have failed to strike. Cassava has a great ability to compensate and plants from replacement cuttings, which lag behind neighbouring plants, do not manage to develop or yield normally.

*Good cuttings are 25 cm long, approximately twice as thick as the medulla and taken from healthy plants.*

## 2.8. Crop maintenance

### Weeds

A distinction must be made between a number of situations, such as farms that have only recently been established and regularly maintained

Fig 28 *Mantsoumba farm: growing cassava (Zaire)*

Fig 29 *Attempting to cut back Stylosanthes with a crosskill roller (Zaire)*

farms, which includes land that was previously under fallow or grassland as well as land on which cassava or an annual crop has been grown.

On cassava plantations that have been newly established after cutting down forest or clearing savanna, the problems of specific weeds often have to be tackled. Where once there was forest, fresh growth of woody plants occurs; where there was formerly savanna, certain perennial *Gramineae* with rhizomes are difficult to eradicate; and on certain light acid soils, farmers are often faced with a *Cyperaceae* problem.

These specific problems must be dealt with as fully as possible when the land is being prepared.

At this stage, the solution generally consists of repeated ploughing or pseudo-ploughing, sometimes combined with applications of herbicide. There is no blanket solution, as each situation is specific. However, it is preferable to use implements with discs for woody plants and tined implements for plants with rhizomes (*Gramineae* or *Cyperaceae*). Herbicidal treatment will sometimes be necessary to deal with particularly stubborn weeds and to obviate the risk of erosion which a large number of mechanical cultivations will entail.

Fallow breaks or grassland that has been grazed over a long period may create similar difficulties on a farm to those that occur after savanna has been cleared.

Green manure crops or grassland consisting of *Stylosanthes* may leave behind in the soil large quantities of seed, the germination of which will be staggered over a very long period of time and which will lead to fresh growth whenever the soil is insufficiently covered by cassava. This new growth, which will be found in old, partly defoliated fields may require manual weeding. However, *Stylosanthes* cover generally smothers any competition and helps keep the land clean. Every cropping system tends to eradicate certain types of weeds and preserve others.

In a rotation consisting of annual crops, the preparatory and maintenance cultivations — repeated at regular intervals — tend to favour slow-growing species. With cassava-based rotations, the interval between cultivations is often greater than a year and there may be long periods at the end of the cycle when the cassava does not provide good soil cover. Such conditions encourage the development of perennial species which may initially not appear to be particularly troublesome, but which must not be allowed to become established. With cassava, weed control must be considered from a multiannual perspective, and not only in terms of the most immediate effects.

## When must action be taken against weeds?

During the first three or four months of its growth, cassava does not cover the soil and weeds tend to proliferate. Right at the beginning, when the cassava is starting to grow, if the young cuttings are covered, even temporarily, by weeds, yields will be substantially reduced. In such a situation, the weeds are competing not only for water, but also for light. If they do not cover the plants, but only the soil or part of the soil, they are competing merely for water and the effect on yields will probably be less serious, but the less frequent the rain, the greater the effect. Maintenance operations during this period must be such that the soil is kept as clean as possible. They must be carried out when the weeds are starting to grow and not once they have become established.

The first weeding should take place immediately after planting, followed by two or three further weedings before the fourth month.

From the fourth month onwards, cassava cover is generally sufficiently dense to prevent any vegetation from growing.

At the end of the cycle, i.e., from 8, 10 or 12 months, depending on the situation, the foliage will become less dense, as a result of a decrease in the size of the leaves and/or a reduction in the number of leaves. Drought or disease may accentuate this phenomenon. Where there is a substantial reduction in the amount of foliage at the end of the cycle, weeds will develop. The effect on yields is far less marked than at the start of the cycle but further weeding may be necessary, in particular to prevent species that are difficult to control from developing among subsequent crops.

## How are weeds to be controlled?

Weeds can be controlled mechanically or chemically or by a combination of both methods.

## Mechanical weeding

The way in which the land is prepared contributes to this method of control.

After planting, the mechanical operations consist of weeding and hoeing. Earthing-up is sometimes preferred, as it increases the volume of loose earth at the foot of the plants and thus encourages root growth. It is also thought that earthing-up may reduce the incidence of lodging, but this has not been proved. Lastly, earthing-up may reduce weed growth along the rows and make manual intervention unnecessary.

Where the cuttings have been planted on the flat and it is decided to use earthing-up as a means of maintaining the crop, an early start must be made, before the superficial lateral roots have reached the middle of the space between the rows, so that they are not destroyed. Lastly, if a herbicide was employed at the time of planting, earthing-up is not advisable, as it will cause the herbicide to be concentrated at the foot of the plants.

Mechanical maintenance operations at the start of the cycle are not always sufficient and further weeding by hand may sometimes by necessary along the rows. It is also possible to combine the application of herbicides with mechanical operations. At the end of the cycle, weed control cannot be done by mechanical means.

## Chemical weeding

*The products*

A large number of herbicides can be used with cassava and new products are being tried out every year. Some are more suited than others to particular weeds.

It is therefore advisable to seek information from research centres or agricultural experimental stations on the products that are available and their characteristics. Here, we shall provide only information of a general nature.

Some products are highly selective with regard to cassava, which means that they are not toxic at the recommended doses or often even at much higher doses. While others are less selective and certain precautions must be taken when using them, they are still employed either for economic reasons or because they provide the only means of destroying certain weeds.

The selectivity or toxicity of certain active ingredients with regard to cassava may also vary according to the age of the plant.

The least toxic active ingredients include alachlor, bifenox, butachlor, chloramben, fluordifen, methazole, nitrofen and trifluralin. Products which are only slightly toxic include butylate, diuron, fluorometuron, metribuzin and prometryn and those whose toxicity varies according to the age of the cassava include paraquat and glyphosate, which are completely selective after the plant is three months old, provided that they do not touch the foliage, while under the same conditions they are toxic to younger cassava plants.

*Chemical control*
Some herbicides may be applied in solid form (microgranules) but normally they are applied in liquid form, diluted in water.

The quantities of liquid to be used depend on how the equipment for applying the herbicide works. With spinning disc-type distributors (low-volume), the quantity of liquid to be provided per hectare varies between 30 and 50 l. With the traditional pressurized equipment, between 400 and 500 l is needed.

Both types of equipment are available for use in conjunction with tractors in the case of mechanized cultivation and for manual cultivation (portable equipment). The low-volume portable equipment is particularly valuable as a labour-saving device in cases where mechanization is not possible. In all cases, whether the equipment has nozzles or discs, whether it is tractor-mounted or carried by the operator, it must be fitted with a deflector shield, to enable the herbicide to be directed at the ground and the foliage to be protected.

In practice, the following recommendations may be made:

(a) in order to keep the soil clean at the start of the cycle:
either apply the herbicide prior to planting by incorporating it in the soil (butylate may be used at a rate of 4 kg/ha of a.i.*);
or apply the herbicide after planting the cassava but before it germinates, the products which may be used being diuron, alachlor, linuron, oxyfluofen, etc. These products may be employed alone or in combination at doses varying, according to the product, from 0.5 to 2 kg/ha of a.i.;

(b) in order to weed under cassava that is growing, direct the applications, scrupulously avoiding contact with the cassava leaves;
the products to be used are:
diuron (2 to 4 kg/ha of a.i.) if the plants less than three months old;
paraquat (0.5 kg/ha of a.i.) or glyphosate (1 kg/ha of a.i.) if the plants are at least three months old.

N.B. It is essential to follow label recommendations for all agrochemicals since formulations and concentrations can change with brands.

## Integrated control
Integrated weed control involves the application of a number of different methods. In a sense, weed control in the case of cassava is always

* active ingredient

integrated in that it employs not only specific weed-eradication techniques but also concern is shown for weed control when the land is being prepared and even when deciding how long the cropping cycles should last and what the rotations should be. We shall limit our recommendations to those that can be made for a combination of mechanical and chemical treatments.

Chemical control alone is not desirable, as it may lead to the creation of a resistant flora. Where a combination of mechanical or manual and chemical weeding is used as the crop is starting to grow, it does not matter which of these control techniques is used first. If the cassava is chemically weeded once and manually weeded once, this is equivalent to weeding by hand three times.

Herbicidal treatment may be combined with the preparatory cultivations in order to reduce the number of subsequent cultivation operations in the event of particularly recalcitrant weeds. A herbicide in granule form may be used.

> *Take effective action against weeds at the start of the growth period, then keep a close watch.*

## 2.9. Protecting the crop against pests and disease

Generally speaking, the insects and mites that are harmful to cassava and the various diseases that may affect it do not justify treating the crop with pesticides. The reasons for this are almost invariably economic, but sometimes also technical. In rare cases, crop protection may require the cuttings to be treated but normally it is general precautions that are called for.

The advice that can be given is summarized below.

### Treating the cuttings

Cuttings that have been properly selected and prepared do not need to be treated. Treatment must, however, be considered in the case of cuttings that are in poor condition but cannot be replaced, in particular cuttings that were prepared too long ago, or ones that have become weak or been damaged by various fungi or insects. Cuttings that come from a neighbouring region must also be treated, in order to reduce the risk of introducing parasites.

A formula used by research stations to treat cassava is as follows:

per hectolitre of solution: 350 g of captafol, 175 g of endosulfan, 35 g of benomyl and 500 g of zinc sulphate.

## Selecting the varieties

It is now possible to obtain from the research stations varieties that are resistant to or tolerant of at least two major diseases, namely mosaic and bacterial blight. These varieties may not, however, be available or not yet have been tried out in one's own region. If these diseases are not particularly virulent in the region concerned, non-resistant varieties may be preferred as their average yields are higher than those of resistant varieties. In such cases, attention must be confined to plants that look healthy when selecting cuttings.

## Selecting healthy cuttings

This matter has already been raised in the section dealing with planting. Here we shall highlight the difficulties that may be encountered at certain times of the year, when the symptoms of the disease may be less apparent than at other times. The angular spots characteristic of bacterial blight develop particularly during the rainy season, and a plant that has been attacked by mosaic virus may not show any signs of the disease on new shoots that develop at a time of strong vegetative growth, even though the stock remains infected. Thus it is not enough simply to assess the plant's state of health at the time when the cuttings are being harvested. Growers must learn to observe the crop all the year round and recognize those areas that are healthiest. In some cases, it will be possible to take cuttings only from those areas, while in others it will be necessary to select healthy cuttings from dubious-looking areas.

## Cultural techniques, fertilizing the crop

Techniques that encourage better drainage of the soil reduce the risk of root rot. Appropriate, balanced fertilizer use is likely to cut down on the incidence of bacterial blight. Finally, good crop management invariably contributes towards improving the state of health of the plants.

## Biological control

It should be noted that direct control of certain insects is not within the means of individual farmers, but a method that is often effective consists of encouraging the development of diseases or parasites that affect the harmful insects in question. This method involves first identifying the diseases or parasites that affect the harmful insect, selecting them, breeding them and releasing them in the infested areas.

Such action can be taken only on a regional scale. Research is under way to develop biological control, particularly of bugs and certain mites.

*The health of a crop may be improved by using selected varieties, choosing healthy cuttings and applying good cultural practices.*
*It is normally uneconomic to use pesticides, except for disinfecting cuttings.*

## 2.10. Harvesting

Harvesting the tubers and removing them from the field is the most important operation of all, not only because this is the ultimate objective of the production process but because it requires a lot of work, organization and speed. It is the most expensive operation. This is because the crop has to be pulled up from the soil, because large tonnages are involved and lastly because it is important to remove the crop rapidly since it does not keep well.

Complete mechanization of all the harvesting operations is the aim of all farms of any size. However, worldwide it is still the exception, for a number of reasons. The principal ones are that a considerable amount of power is required to lift the tubers, the fact that labour is still cheap in the cassava-producing countries and that it is generally possible to spread harvesting throughout the year, which makes it possible to use permanent labour and avoid the uncertainties encountered with other crops when use is made of seasonal workers.

Harvesting may be divided into a number of distinct operations: cutting the stems, lifting the tubers and shaking them to get rid of the earth, loading and transportation.

### Cutting the stems

The stems may be used in three ways:

(a) they can provide cuttings: the stems are cut by hand, pruned, sorted and tied together in bundles. This aspect has been dealt with earlier (preparing the cuttings);

(b) they may be returned to the land: here again, details have already been given in the section on field preparation, although the following remarks may be added:

- *manual treatment* The stems that have been cut by hand may be left in the field in windrows. This practice is possible if cassava is not planted again immediately but if a long-term green manure crop is sown. The green manure will cover the windrows and encourage them to break down. However, this approach is likely to make it more difficult to work the green manure into the soil using mechanical means. Stems that have been cut by hand and heaped into windrows may be burned after partial drying. This practice soon provides a clean field but it causes losses of organic matter and nitrogen.

- *mechanical treatment* We have seen that mechanically shredding the stems using rotary choppers is the most satisfactory solution;

(c) they may be used outside the farm.

In certain projects for the agro-industrial exploitation of cassava, it is planned to use the stems either as a fuel and source of energy in the processing industries, or as a filler and source of protein in the manufacture of animal feed which also uses the roots. These projects have hitherto normally involved cutting the stems by hand. Mechanical cutting with shredding would appear to be possible with maize forage harvester-type equipment. Trials recently carried out in Ivory Coast with this type of equipment but aimed at merely getting rid of the aerial parts suggest that this is possible where the mass of green matter does not exceed 20 t/ha, which corresponds to an equivalent or slightly higher production of roots.

## Harvesting the tubers

### Harvesting by hand

Harvesting by hand is easy only if the soils are fairly sandy. Then, the picker, after cutting the stem 10 or 15 cm above the neck, pulls on the piece of stem that has been left and fairly easily lifts up all the tuberous roots, although sometimes the tubers break off and an implement such as the *daba* (a type of African hoe) is used to retrieve them. On soils that are more clayey, a *daba* must be used to lift the whole of the plant and, if

Fig 30 *Mantsoumba, a three-year-old plant which has just been lifted, showing tubers of different ages (Zaire)*

Fig 3.1 *Area for drying cassava (Ghana)*

the soil is also dry, harvesting by hand may be a lengthy and difficult business. Various instruments have been devised to facilitate harvesting, such as a form of pliers which grip the base of the stem and rest against the soil to lever the plant out. These tools do not provide any real improvement.

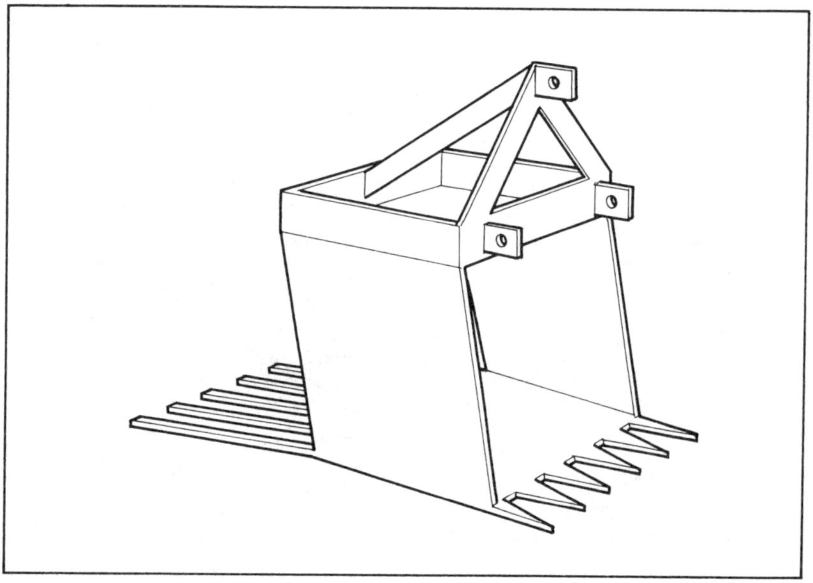

Fig 32 *Diagram of a tractor-mounted cassava harvester*

## Semi-mechanized harvesting

Where conditions are difficult (heavy or dry land), manual harvesting can be facilitated by drawing along the rows a trailed implement which divides the soil. The kind of implement that can be used is a plough with the mouldboard removed or a subsoiler, possibly fitted with a wing. Such solutions have proved satisfactory in Madagascar.

## Mechanical harvesting

Mechanical harvesting is not yet very widespread. Numerous equipment

Fig 33 *Cassava plant at harvesting*

Fig 34 *Mechanical harvesting*

manufacturers, particularly in France, the United Kingdom and Brazil are working on this problem.

Cassava harvesters work on the principle of cutting a slice of earth about 30 cm deep and 70 to 80 cm wide, lifting it and then separating the roots from the soil adhering to them.

Cassava harvesters therefore contain at the front a strong share for lifting and, at the rear, a vibrating device for separating the tubers from the soil and stones. Mechanical harvesting of cassava requires very powerful machinery (at least 100 hp per row).

## Loading, transport and sorting

Loading equipment is being studied in the same way as harvesting equipment. What is needed is a machine that will follow the harvester, collect the tubers on a conveyor and deliver them to a trailer.

Usually, tubers that are harvested by hand or semi-mechanically are loaded on to trailers by hand.

Lastly, harvesting involves one final operation – which is always done by hand, either in the field or in a collecting area (entrance to the factory) – consisting of separating the roots from one another and removing their peduncle and the fragments of stem. This operation is often referred to as 'topping' and is used as an occasion for sorting the crop and removing unwanted tubers, depending on the purpose for which they are to be used.

*Harvesting is the most expensive operation and requires excellent organization.*

Fig 35 *Mechanically loading cassava after harvesting*

## 2.11. *Labour requirement*

Cassava is sometimes regarded as a crop which requires little in the way of labour or, on the contrary, as very labour-intensive, depending on whether you are referring to a very extensive traditional system or a partly improved system or, again, according to whether or not this labour is related to the high output of useful matter which can be obtained.

The amounts of labour required under different production conditions and with different cropping methods may vary considerably. To give some idea, the following indications can be given:

### Traditional manual cultivation

*Clearing the land*: 30 to 40 man days' work/hectare. Clearing old savanna takes as long as and often longer than clearing forest.

*Preparing the soil*: very variable, 15 man days' work/hectare on average. After the land has been cleared, there is normally no preparatory cultivation; the cuttings are planted directly with a little hoeing. A certain amount of preparatory cultivation is carried out only on fields that have already been cultivated.

*Taking and planting the cuttings*: 20 man days' work/hectare. As a rough guide, 6 man days' work/hectare is needed to take and prepare the cuttings and 14 man days' work/hectare to plant them.

*Maintenance*: 5 to 30 man days' work/hectare. Maintenance is very limited and even non-existent after forest has been cleared. A large amount of maintenance is required on fields that have been previously cultivated but the amount of work that is necessary is seldom carried out.

*Harvesting*: 20 to 25 man days' work/hectare. The amount of work involved in harvesting of course varies considerably according to the yield of the crop but is not simply proportional to yield. The amounts of work quoted here are for yields of 8 to 10 tonnes/ha, which are the normal yields under traditional systems. To harvest a crop of 20 t/ha will take more than 40 man days' work/ha.

*Transport*: normally 4 man days' work per km per tonne should be allowed for. With traditional manual cultivation, the total labour input per hectare is between 90 and 130 man days, which, for the yields normally obtained, is equivalent to 8 to 15 man days' work per tonne of roots produced.

## Improved cultivation, mechanized or partly mechanized

*Preparing the land:*

- ploughing: three-disc plough, 90–100 hp tractor, 2 hours/ha;
- working down of ploughed land or cultivation that is lighter than ploughing: heavy disc harrow and 90–100 hp tractor, 1.5 hours/ha;
- working down of ploughed land or cultivation with rigid-tined implements to eradicate *Gramineae* with rhizomes: 90 to 100 hp tractor (more than one pass is frequently required), 6 to 9 hours/ha.

*Planting:*

- taking cuttings: work done by hand — 4 to 5 man days' work/hectare;
- taking cuttings mechanically: 1 hour/ha.
- transporting: tractor and trailer, 1 hour/ha;
- planting with four-row planter and tractor, 1 hour/ha.

*Fertilizing:*

- spinner/broadcaster and 60 hp tractor, 0.5 to 1 hour/ha.

*Maintenance:*

- earthing-up: one pass with a ridging-plough and 65 hp tractor, 2 hours/ha;
- maintenance by hand along the rows, 6 to 8 man days/ha.

*Harvesting:*

- chopping of the aerial parts: rotary chopper and 100 hp tractor, 3 hours/ha;
- mechanical harvesting: single-row harvester and 100 hp tractor, 2.5 to 3 hours/ha.

Although so far we have not had any real experience of a fully mechanized farm, it can be estimated that such a farm, equipped with 60 to 70 hp tractors for the light work and 100 to 110 hp tractors for the heavier work, specific cultivating implements and trailers, would require per hectare:

25 to 30 hours' work with tractors, including transportation within the farm;
40 to 60 hours' labour;
250 to 300 litres of fuel;
400 to 600 kg of fertilizer;
herbicides, fungicides for disinfecting the cuttings and various other products.

Normally, the modernized farms are only partly mechanized, only the preparatory cultivations, some of the maintenance work and the transportation being mechanized. Under these conditions, 20 man hours' mechanized work and 500 man hours' manual labour may be needed.

One step lower down the scale, where only the preparatory cultivations are mechanized, 5 to 10 hours of mechanical work per hectare may be needed, depending on the amount of weeds present. With such a system, the amount of manual labour required is of the order of 100 man days/ha, as much as with completely manual traditional cultivation. The main difference is in yields, which, with an improved system, may be twice as high as with traditional cultivation.

# 3 Cassava products

The purpose of this book is not to deal in detail with techniques for processing cassava or with the uses of the various products that may be obtained from cassava. A very brief survey of these matters follows:

## 3.1. *Traditional foodstuffs*

Cassava is regarded as one of the principal plants of use to man because of the important part it plays as a food.

### Food characteristics

It is of course the tuberous roots that are most important as a foodstuff, but it should be noted that the leaves may also be eaten and that they possess nutritional characteristics which are complementary to those of the roots. The following average figures may be given:

**Table 3**

| | | **Complete roots** | **Peel** 8 to 15% of the root | **Stem** | **Leaves** |
|---|---|---|---|---|---|
| **Dry matter** (DM) | % | 35 | 30 | 40 | 15 |
| **Glucids** (mainly starch) | % DM | 89 | 75 | 91 | 41 |
| **Lipids** (fats) | % DM | 1 | 2 | 0.5 | 6 |
| **Protein** | % DM | 2.5 | 4 | 2 | 25 |
| **Fibre** | % DM | 4.5 | 12 | 4 | 20 |
| **Ash** | % DM | 3 | 5 | 2.5 | 8 |

It can be seen that the food value of the tubers lies in their high starch content, the other categories, such as fats and proteins, being present only in small amounts.

The products derived from cassava are high-energy foods of excellent quality in that cassava starch is particularly good in terms of digestibility. It is not a complete food. It must be supplemented with fats and above all the proteins that are necessary for proper nutrition. Indeed Kwashinorkor is a protein deficiency disorder associated with excessive cassava and other starchy roots in the diet.

Note that the leaves, whose dry matter contains approximately 25% protein is an excellent complement to the roots. That is probably why cassava leaves are widely consumed in countries where the tubers are the staple food.

Lastly, it must not be forgotten that cassava contains a glucoside which, as it breaks down, produces hydrocyanic acid. The processing or cooking which the tubers or leaves undergo prior to being consumed eliminate virtually all these substances to a sufficient extent to prevent poisoning.

**Dried cassava**

Drying the roots is the simplest method of preserving cassava.

The tuberous roots, either peeled or unpeeled, are cut up into chips (*cossettes*) and placed on mats, directly on the ground or on a cement-rendered surface to dry in the sun, which takes two to three days.

The products obtained from whole roots are often sold for animal feed.

For human consumption, it is chips from peeled roots that tend to be used. These chips are normally crushed, sieved and reduced to a flour. This is the Brazilian *farinha de raspa* or the Indonesian *glapeck*.

The flour is used to make porridges, doughs or bread. It may also be mixed with wheat flour, the proportion being around 20%, without impairing bread-making characteristics.

**Cassava steeped in water**

The roots, either peeled or unpeeled, are immersed in water, preferably slowly running water, for three to five days. After this treatment, the peel (if it has not previously been removed) and the fibres come off very easily; they are removed and the pulp is dried in the sun. The dried fragments thus obtained are pounded into a flour, which is very white and possesses a distinctive, pleasant taste which is due to the fermentation that takes place during steeping. This is the Brazilian *farinha de agua* or the Central African *farine* or *fufu*. Cassava may also be made into 'cassava bread' and 'cassava sticks' or 'chickwangues'. Preparation is complicated, consisting of repeated working or kneading, preliminary cooking to produce a dough, further kneading, shaping into loaves or sticks, wrapping in leaves and further cooking.

Cassava loaves and sticks will keep for eight to ten days.

**Grated cassava products**

The fresh roots are peeled, then grated. The grated pulp is normally left to ferment, sometimes with leaven from a previous batch. The period of fermentation varies according to country (one to three days). The pulp is then wrung dry by compression, then broken up and finally cooked on heated plates. In some cases, the broken-up pulp is simply dried in the sun. The product obtained is similar to semolina and it is used in the same way as couscous.

This processing technique is used in many cassava-producing countries. The principle is usually the same, but each region has its own tradition with regard to the practical details and the products obtained each have their own individual characteristics. These are the Brazilian *farinha de mesa*, the Guyanese *kwako* or *couac*, the West African *gari* or *atieké*, and the Angolan *musseques* and *malanje* flours.

**The leaves**

The tips of the branches and the young leaves are consumed in various culinary preparations. Boiling the leaves, which is the first stage in these preparations, is necessary in order to eliminate the hydrocyanic acid.

## 3.2. Industrial products

### Pellets

The industrial product which is currently most important in terms of the volume traded is cassava pellets. These are manufactured as follows: the tubers, normally unpeeled, are cut up, using root cutters, into chips, which are dried in the sun on concrete. This stage of the manufacturing process may take place on the farm or in the village. The pellets are then produced in the factory by compressing the chips against grids with an 8 to 10 mm diameter mesh.

Compared with dried chips, the pellets obtained in this way have the advantage of being of uniform size (the meal always present in the chips has been agglomerated) and above all that of having a higher apparent density, which facilitates transportation and marketing. The pellets are used, principally in Europe, for manufacturing animal feed.

### Starch

Starch is virtually pure carbohydrate. It is used for various purposes in the food industries (sweetened products, thickeners, making tapioca, etc.), in the textile, paper and other industries, but cassava starch is often competing with that derived from other plants (maize, potatoes).

Fig 36 *Traditional preparation of gari, sifting the pulp (from which the liquid has been squeezed out), prior to cooking*

The starch is produced in large factories, where the sequence of operations can be summarized as follows: washing the tubers, peeling, grating, extracting the starch, washing the starch, refining and drying. Various procedures may be used to extract the starch and they are generally based on filtration and centrifugation.

## Tapioca

Tapioca is a foodstuff prepared from cassava starch. The starch, which is still moist after being refined, is remoistened, passed through sieves, gelatinized by heating and finally dried.

## Other industrial products

Other industrial products are possibilities rather than realities. These possibilities depend on economic factors and/or further research and technical development.

The following should be mentioned:

(a) *sweeteners*, particularly syrups rich in fructose, which are widely used in the manufacture of sweetened drinks. These syrups are made from starch, mainly maize starch. They could be produced from cassava starch in the cassava-producing countries;

(b) *alcohol*. For some time, interest has been shown in numerous countries in the manufacture of industrial ethyl alcohol from agricultural products rich in carbohydrates. With the oil crisis, there has been renewed interest, notably in Brazil, in the manufacture of ethanol from cassava;

(c) *fermented products*. Various forms of fermentation are being studied with a view to preserving cassava or obtaining protein-enriched products. In the former case, this involves techniques for ensiling

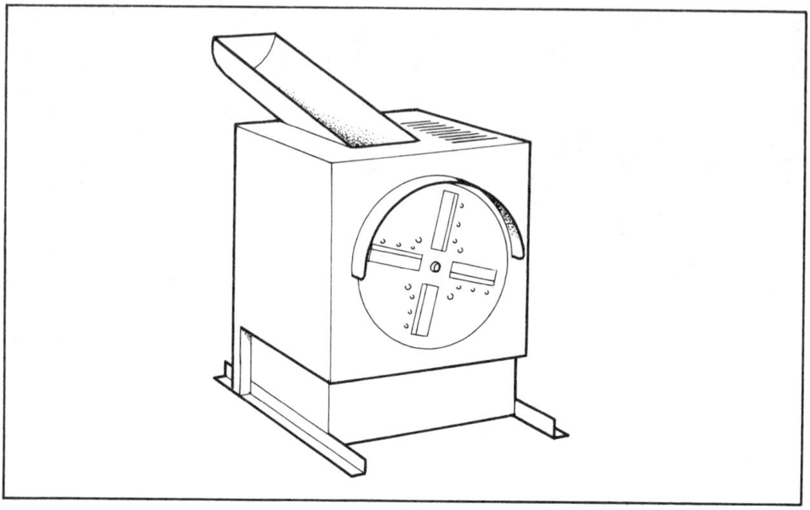

Fig 37 *Malaysian-type chipping machine*

Fig 38 *Preparing pieces of dried cassava in Madagascar*

pulped roots, to which could be added the aerial parts, also shredded. The aim would be to produce an animal feed to be used on the spot, which would not need drying and which woud keep well. This would involve the use of appropriate micro-organisms to convert some of the root starch into protein, in order to obtain a product containing 15 to 20% protein which could be used in animal feed or even for human consumption.

# Appendix

## *1. Cropping plans, rotations and patterns*

Once a farms starts to modernize its operations, reaches a certain size and has to meet a demand which must be satisfied on a regular basis, it is essential to study the cropping plan thoroughly. The examples given here correspond to certain situations that may be encountered and provide a method that may be used in other cases.

The following situations are examined:

**1.1** harvesting and planting take place all the year round;
**1.2** harvesting and planting take place at certain times of the year only;
**1.3** harvesting is confined to a short period and there are two periods for planting;
**1.4** harvesting occurs throughout the year but there is only one period for planting;
**1.5** harvesting occurs throughout the year but there are two periods for planting.

The cassava crops, in terms of their growth periods and where they fit into the farming year, will be referred to as 'cycles' and designated by the letters A,B,C,D etc. They will be represented in the rotation tables by the same upper case letter for the preparation and start of the cycles in question and by the corresponding lower case letter a,b,c,d, etc. for the harvests. This distinction is necessary as a cycle is seldom contained within one and the same year.

The key to the symbols used in the charts is as follows:

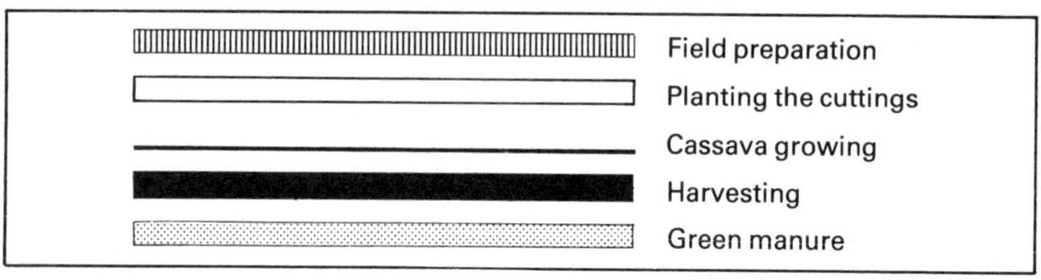

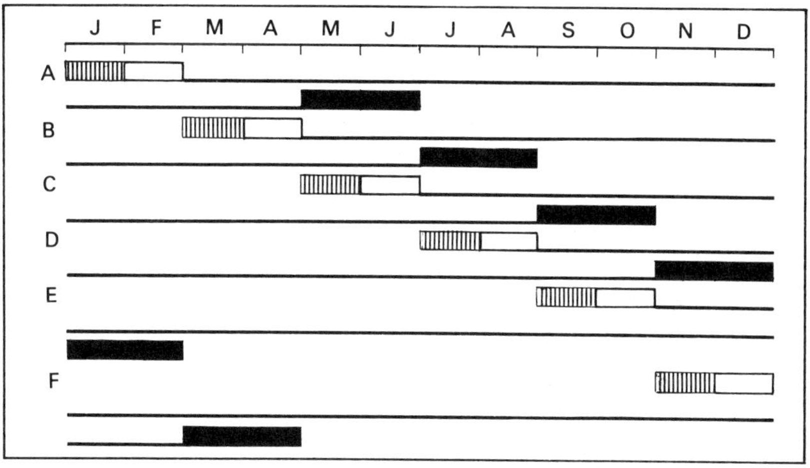


Fig A.1 *Possible cycles for 15 to 16 month cassava with no constraints with regard to timing of planting and harvesting*

## 1.1. Harvesting and the planting of cuttings take place all the year round

This situation may be encountered in a region where there are no marked variations in climate and where production is intended for an industry or local market whose demand is continuous.

### Cycle and cropping schedules

A single cropping period may be selected for all cycles which differ only in terms of where they fit into the schedule. Here, we have taken a period of 18 months, corresponding to cassava which is harvested when the crop is 15 to 16 months old, and six cycles with harvesting spread over two months. These cycles are shown schematically in Figure A.1.

### Rotation and patterns

Three cases are possible: continuous cropping of cassava (although that is not desirable), a green manure break between each cassava crop and a green manure break every two cassava crops.

*Continuous cropping of cassava*

It can be seen from the chart above that, after harvesting cycle A, the land can be prepared immediately for cycle D, after which one goes back to cycle A again. The same applies to the sequence of cycles BE and CF. In this case, the farm will follow three separate three-year rotations (Fig A.2).

Thus the land will be divided up into three blocks, each assigned to

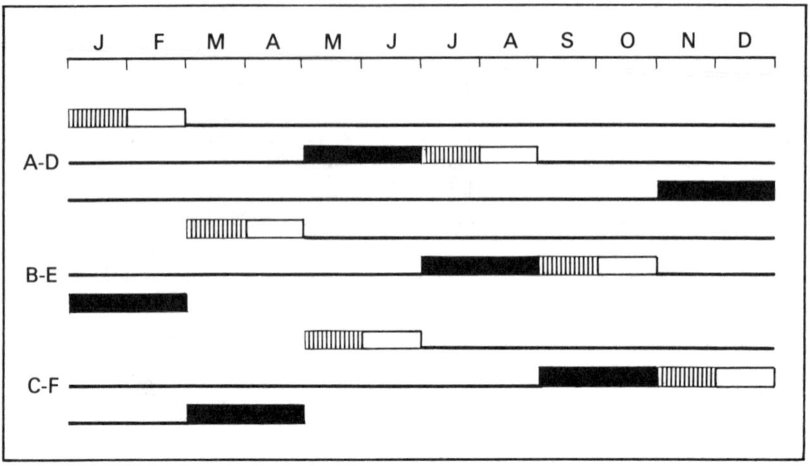


Fig A.2 *18-month continuous cassava rotations*

one of the rotations, and each block will be divided into as many fields as there are years in the rotation. The rotation can be presented in schematic form as follows:

It can be seen that the system is fully established in the third year, with harvests a,b,c,d,e and f on six plots (Fig A.3), and that these harvests cover the whole year, i.e., e in January-February, f in March-April, a in May-June, b in July-August, c in September-October and d in November-December.

It should also be noted that the length of a rotation is a multiple of the length of the cycle. Here, two cycles make a complete number of years, so that with six cycles, three rotations are needed. A different cycle length could have led to two rotations or a single rotation. It is necessary to weigh up the advantages and disadvantages of long- or short-rotation systems, some being quicker to introduce than others.

| | Block I | | | Block II | | | Block III | | |
|---|---|---|---|---|---|---|---|---|---|
| Field | 1 | 2 | 3 | 1 | 2 | 3 | 1 | 2 | 3 |
| Year 1 | A | D | — | B | E | — | C | F | — |
| Year 2 | aD | d | A | bE | | B | eF | | C |
| Year 3 | d | A | aD | | eB | bE | | fC | cF |

Fig A.3 *18-month continuous cassava rotations*

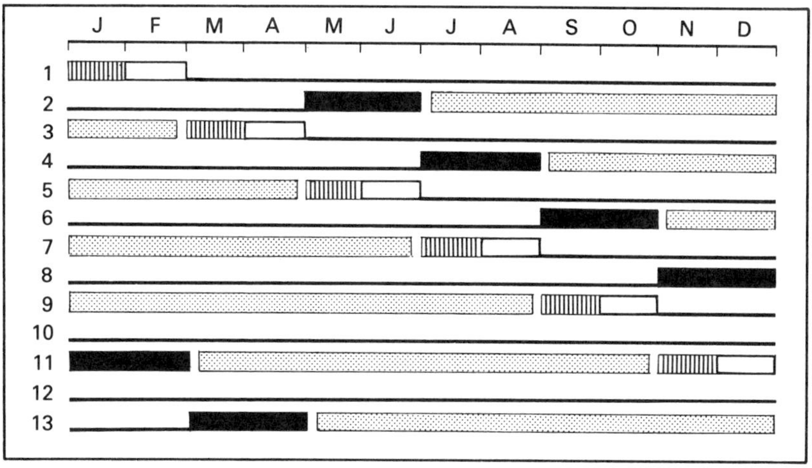


Fig A.4 *Rotating 18-month cassava with an 8-month green manure break*

*Cassava/green manure*

If a green manure (gm) is sown after a cassava crop, it can occupy the ground for four, six or eight months, for instance. In the first case, the

| Year \ Field | 1 | 2 | 3 | 4 | 5 | 6 | 7 | 8 | 9 | 10 | 11 | 12 | 13 |
|---|---|---|---|---|---|---|---|---|---|---|---|---|---|
| 1 | A | B | C | D | E | F | | | | | | | |
| 2 | a gm | b | c | d | | | A | B | C | D | E | F | |
| 3 | B | C | D | E | eF | f | a | b | c | d | | | A |
| 4 | b gm | c | d | | | A | B | C | D | E | eF | f | a |
| 5 | C | D | E | eF | f | a | b | c | d | | | A | B |
| 6 | c gm | d | | | A | B | C | D | E | eF | f | a | b |
| 7 | D | E | eF | f | a | b | c | d | | | A | B | C |
| 8 | d | | | A | B | C | D | E | eF | f | a | b | c |
| 9 | E | eF | f | a | b | c | d | | | A | B | C | D |
| 10 | | | A | B | C | D | E | eF | f | a | b | c | d |
| 11 | egmF | f | a | b | c | d | | | A | B | C | D | E |
| 12 | | A | B | C | D | E | eF | f | a | b | c | d | |
| 13 | f gm | a | b | c | d | | | A | B | C | D | E | eF |

Fig A.5 *Rotating 18-month cassava with an 8-month green manure break (the sowing of the green manure is shown only in field 1)*

sequence would be A — gm — F — gm — E — gm — D — gm — C — gm — B – gm. This rotation would last 11 years (the length of a cassava/green manure cycle is 22 months; the six cassava/green manure cycles therefore together cover 132 months, i.e., 11 years). In the second case, there would be six distinct breaks, i.e., A — gm — A — gm, etc., B — gm — B — gm, etc., C — gm, C — gm, etc. In the third case, there would again be a single 13-year rotation (26 × 6 = 12 × 13). This last case is shown in schematic form in Figure A.4.

In the last case, the rotation will consist of 13 fields, as shown in the table on page 69 (Fig A.5):

*Cassava/cassava/green manure*
Still with an 8-month green manure crop, the rotation A — D — gm — E — B — gm — C — F — gm lasts 11 years. It is shown in Fig A.6. The breakdown by field is given in Fig A.7.

### 1.2. Harvesting is confined to a short period and there is only one period for planting cuttings

These conditions may apply in the case of a farm which intends to limit production to the dry season, so that the roots can be dried in the open air.

**Schedule and cropping cycles**
The period that is favourable for the planting of cuttings normally occurs immediately after the dry season.

In practice, there are only two possible types of cycles for which the

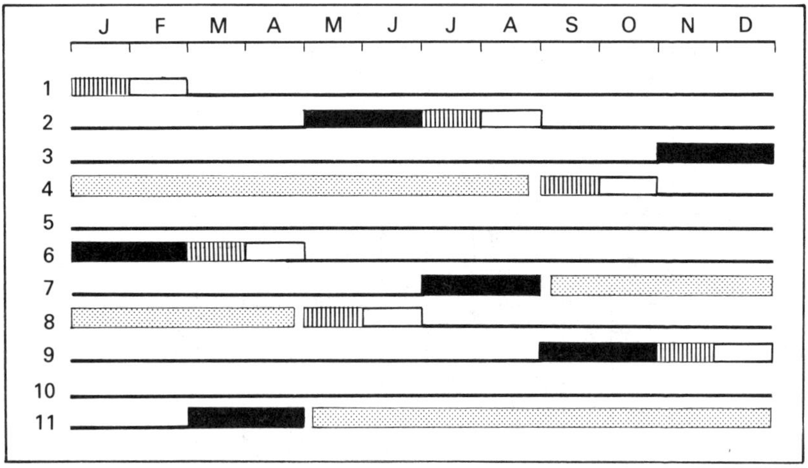


Fig A.6 *Rotating 18-month cassava with an 8-month green manure break every two crops*

| Year \ Field | 1 | 2 | 3 | 4 | 5 | 6 | 7 | 8 | 9 | 10 | 11 |
|---|---|---|---|---|---|---|---|---|---|---|---|
| 1 | A | D | | E | | B | | C | F | | |
| 2 | aD | d | E | | B | b | C | cF | | | A |
| 3 | d | E | | eB | b | C | cF | | f | A | aD |
| 4 | gm E | | eB | b | C | cF | | f | A | aD | d |
| 5 | | eB | b | C | F | | f | A | aD | d | E |
| 6 | eB | b | C | cF | | f | A | aD | d | E | |
| 7 | b gm | C | cF | | f | A | aD | d | E | | eB |
| 8 | C | cF | | f | A | aD | d | E | | eB | b |
| 9 | cF | | f | A | aD | d | E | | eB | b | C |
| 10 | | f | A | aD | d | E | | eB | b | C | cF |
| 11 | f gm | | aD | d | E | | eB | b | C | cF | |

Fig A.7 *Rotating 18-month cassava with an 8-month green manure break every two crops (here again, the sowing of the green manure is shown only for field 1)*

age at which the cassava is harvested differs by 12 months.

In the example chosen (a four-month dry season and four months of harvesting), either an 8 to 12-month cycle (A) or a 20 to 24-month cycle (B) may be adopted (Fig A.8).

## Rotation and patterns

We shall take as an example a rotation of cassava/cassava/green manure; the green manure break may last 11 months. The rotations are as follows (Fig A.9):

Cycles A and B could be combined in the same field rotation; cycle A would be harvested at the end, and cycle B at the beginning of the dry season. The rotation would then last eight years.

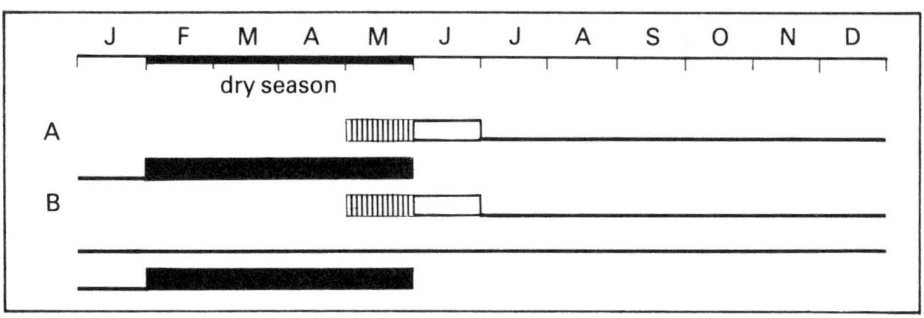


Fig A.8 *Possible cycles for a single period for the planting of cuttings and a single harvesting period*

| Year | Field 1 | Field 2 | Field 3 |
|---|---|---|---|
| 1 | A | A | gm |
| 2 | aA | a gm | A |
| 3 | a gm | A | aA |

| Year | Field 1 | Field 2 | Field 3 | Field 4 | Field 5 |
|---|---|---|---|---|---|
| 1 | B | — | B | — | gm |
| 2 | — | B | — | gm | B |
| 3 | bB | — | b gm | B | — |
| 4 | — | b gm | B | — | b gm |
| 5 | b gm | B | — | bB | B |

Fig A.9 *Cassava/cassava/green manure rotations for 12- and 24-month cycles*

## 1.3. Harvesting is confined to a short period and there are two periods for planting cuttings

This case occurs when, in addition to the main dry season, which is when the crop is to be harvested, there is a shorter, secondary dry season. The second period for planting the cuttings would then be immediately after that secondary dry season.

However, even if there is only one dry season, it may be advantageous to spread the planting work over two separate periods.

Generally speaking, if the growth period is the same, there will be one period that will be superior to the other for planting the cuttings. The cassava enterprise is therefore organized in such a way that the crops planted during the less favourable period are harvested later than the others.

### Schedules and cycles

In the light of that remark, the two cropping cycles could therefore be schematically represented as follows:

With cycle A, the cassava is harvested between 10 and 11 months, with cycle B between 15 and 16 months.

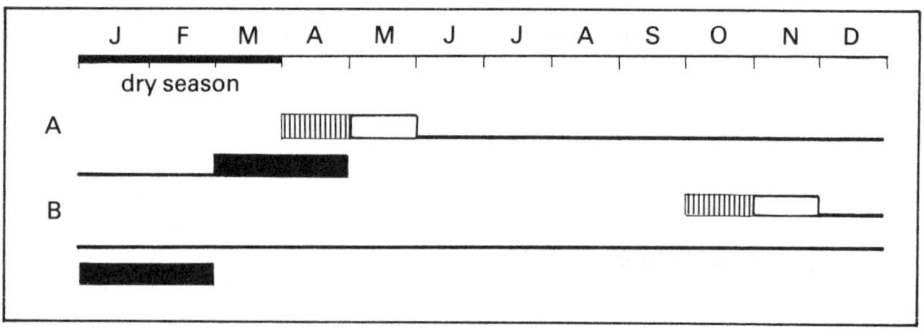


Fig A.10 *Cassava cycles in the case of one harvesting season and two planting periods*

**Rotation and patterns**

This system is well suited to a cassava/cassava/green manure rotation in the order B — A — gm, the green manure being a short-cycle crop.

### 1.4. Harvesting occurs throughout the year but there is only one period for planting cuttings

In this case, the cassava will have to be harvested with an age gap of 10 to 11 months between the youngest and oldest plants. This will inevitably lead to considerable differences in yield according to which plots are harvested and to fluctuations in output which may be difficult for the processing industry to absorb.

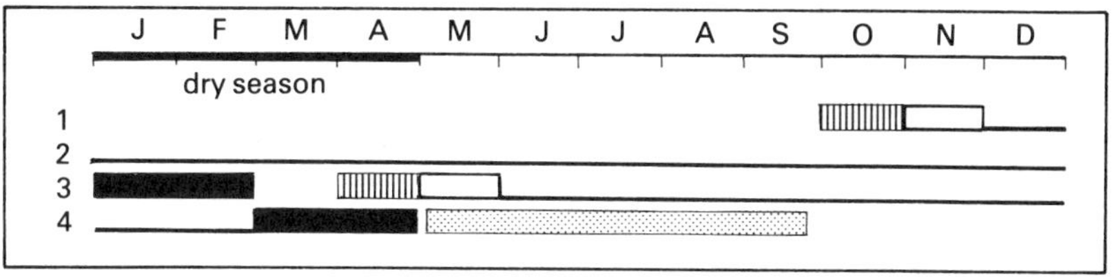


Fig A.11 *Cassava/cassava/green manure rotation, a long cycle alternating with a short cycle*

| | Field 1 | 2 | 3 |
|---|---|---|---|
| Year 1 | B | A | |
| 2 | | a gm | A |
| 3 | bA | B | a gm |

Fig A.12 *Cassava/cassava/green manure rotation, a long cycle alternating with a short cycle*

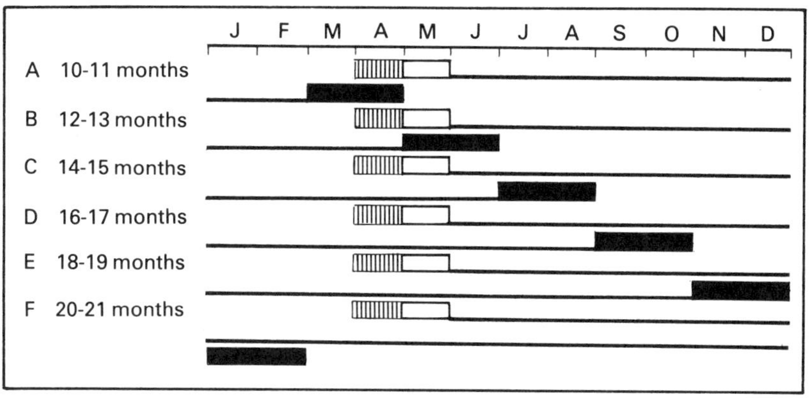


Fig A.13 *Range of cycles with only one period for the planting of cuttings, enabling all-the-year-round harvesting*

## Schedule and cropping cycles

The range of cycles in Fig A.13 could, for instance, be possible:

## Rotation and patterns

For a continuous sequence of these cycles, the rotation would last 12 years but there would be fairly long periods without a crop after the shortest cycles. These periods can be used for a green manure crop. The rotation would then involve a short cycle after a long cycle..

The rotation is as follows:

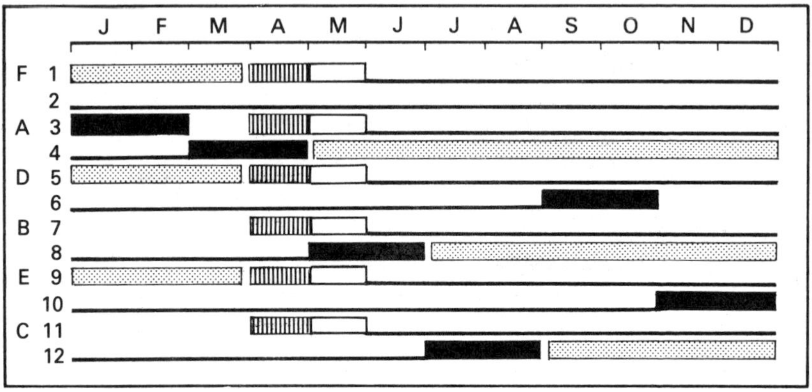


Fig A.14 *Cassava/cassava/green manure rotation in the case of one period for planting the cuttings and harvesting all the year round*

| Year \ Field | 1 | 2 | 3 | 4 | 5 | 6 | 7 | 8 | 9 | 10 | 11 | 12 |
|---|---|---|---|---|---|---|---|---|---|---|---|---|
| 1 | | F | | A | | D | | B | | E | | C |
| 2 | F | | fA | a gm | D | d | B | b gm | E | e | C | c gm |
| 3 | | fA | a gm | D | d | B | b gm | E | e | C | c gm | F |
| 4 | fA | a gm | D | d | B | b gm | E | e | C | c gm | F | |
| 5 | a | D | d | B | b gm | E | e | C | c gm | F | | fA |
| 6 | D | d | B | b gm | E | e | C | c gm | F | | fA | a gm |
| 7 | d | B | b gm | E | e | C | c gm | F | | fA | a gm | D |
| 8 | B | b gm | E | e | C | c gm | F | | fA | a gm | D | d |
| 9 | b | E | e | C | c gm | F | | fA | a gm | D | d | B |
| 10 | E | e | C | c gm | F | | fA | a gm | D | d | B | b gm |
| 11 | e | C | c gm | F | | fA | a gm | D | d | B | b gm | E |
| 12 | C | c gm | F | | fA | a gm | D | d | B | b gm | E | e |

Fig A.15 *Cassava/cassava/green manure rotation in the case of a single period for the planting of cuttings and harvesting all the year round*

## 1.5. Harvesting occurs throughout the year and there are two periods for planting cuttings

This final example corresponds to one of the most frequently encountered situations. Figure A.16 shows a range of possible cycles. Figures A.17 and A.18 provide a schematic representation of the rotations when a green manure break is inserted after two cassava crops.

Note that there is a slight practical difficulty with regard to the preparations for cycle F, which must be carried out during the final stages of harvesting cycle B (9th year field 1, 8th year field 2, etc.).

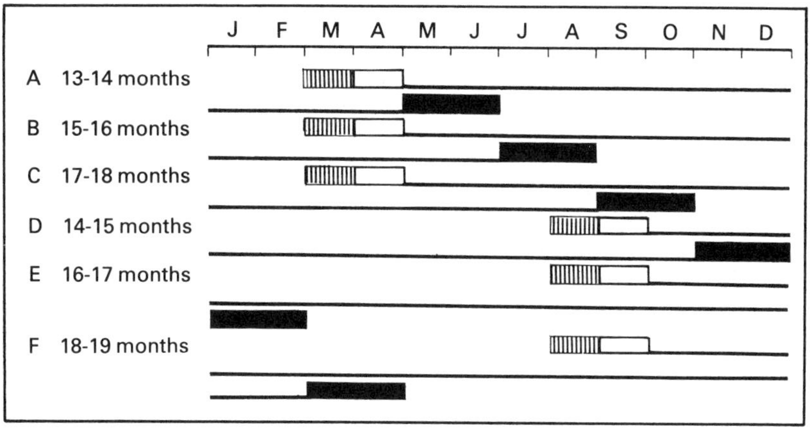


Fig A.16 *Range of cycles when there are two periods for planting the cuttings and continuous harvesting*

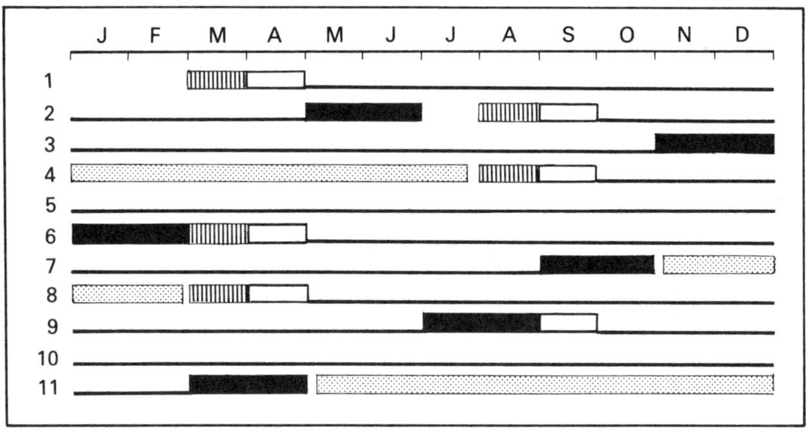


Fig A.17 *Cassava/cassava/green manure rotation with two periods for planting the cuttings and harvesting spread throughout the year*

| Year \ Field | 1 | 2 | 3 | 4 | 5 | 6 | 7 | 8 | 9 | 10 | 11 |
|---|---|---|---|---|---|---|---|---|---|---|---|
| 1 | A | D | | E | | B | F | | | | |
| 2 | aD | d | E | | C | c | B | bF | | | A |
| 3 | d | E | | eC | c | B | bF | | f | A | aD |
| 4 | E | | eC | c | B | bF | | f | A | aD | d |
| 5 | | eC | c | B | bF | | f | A | aD | d | E |
| 6 | eC | c | B | bF | | f | A | aD | d | E | |
| 7 | c | B | bF | | f | A | aD | d | E | | eC |
| 8 | B | bF | | f | A | aD | d | E | | eC | c |
| 9 | bF | | f | A | aD | d | E | | eC | c | B |
| 10 | | f | A | aD | d | E | | eC | c | B | bF |
| 11 | f | A | aD | d | E | | eC | c | B | bF | |

Fig. A.18 *Cassava/cassava/green manure rotation with two periods for planting the cuttings ahd harvesting spread throughout the year*

## 2. *New cassava varieties*

Nigeria has adopted three improved cassava varieties bred at the International Institute of Tropical Agriculture, Ibadan, Nigeria (IITA) after extensive testing in a wide range of environments. These varieties are TMS 30572, TMS 4(2)1425 and TMS 91934. Tuber dry matter yields were highest for TMS 91934 at Ibadan and for TMS 4(2)1425 at Onne (see table below).

**Table: Tuber dry matter yields (t/ha) 1984–1986**

| | **Ibadan** | | | | **Onne** | | | | **Two-location mean** |
|---|---|---|---|---|---|---|---|---|---|
| | 1984 | 1985 | 1986 | **Mean** | 1984 | 1985 | 1986 | **Mean** | |
| **TMS 91934** | 11.0 | 5.6 | 9.8 | 8.8 | 5.1 | 6.3 | 6.0 | 5.8 | 7.3 |
| **TMS 4(2)1425** | 6.6 | 6.0 | 7.9 | 6.8 | 7.6 | 5.4 | 7.4 | 6.8 | 6.8 |
| **TMS 30572** | 7.3 | 5.7 | 6.7 | 6.6 | 4.3 | 5.6 | 5.4 | 5.1 | 5.9 |

In most areas the improved varieties outyielded the local varieties from 43 to 147%. The mean tuber yields for TMS 4(2)1425, TMS 30572, TMS 91934 and 'local varieties' across the 7 or 8 locations for three years were 20.7, 21.7, 24.2 and 10.6 t/ha respectively.

Apart from higher tuber yields the varieties showed high levels of resistance to cassava mosaic virus, bacterial blight and anthracnose disease. They were also resistant to mealybug and green spider mite. TMS 4(2)1425 seemed to be well adapted to different ecological conditions, was high yielding, had low cyanide levels and was poundable after boiling.

The quality of processed food (such as gari, lafun and fufu) made from the improved varieties was rated as good or very good.

# Index